MONEY
ONLINE FORM

Discover The Different Ways Of Making Money Online

Emil Collins

Table of Contents

Just For You!

A FREE GIFT TO OUR READERS

Ready-to-use templates that you can download and start mind mapping with right away!

Visit this Link

https://infobusiness001us.activehosted.com/f/1

Introduction

With an increasing number of people beginning to work from home as a result of the Global Pandemic, the possibilities and opportunities presented by the Online Marketplace have never been greater! In fact, the online Marketplace is essentially booming. It now is a better time than ever to jump on board and take the first step toward establishing your very own empire.

Just to give you an idea of how things are going, an estimated 1.8 billion people were found to purchase their goods online in the year 2018, and the sales were expected to exceed more than 4.8 trillion by now!

At the moment, 77% of the entire Internet Population read blogs, so this has turned into an amazing source of income as well. If you are thinking of Affiliate marketing, the value of the Affiliate Marketing sector in the U.S alone stands at above 7 billion dollars and rising!

And if you are talking about Freelancing, then you should know that almost 43% of the entire workforce of the U.S are Freelancers!

So, yes, the prospects in this sector are huge. But at the same time, the competition is intense as well.

Nowadays, with physical office-based jobs becoming so volatile, an increasing number of people are facing the dread of poverty or living on a very limited day-to-day paycheck. It is becoming increasingly difficult for ordinary people, such as students and entry-level employees, to find work that will allow them to pay their bills, support their families, and live a normal, healthy life.

And even if by a stroke of luck, you are able to get into a job, in most cases, the work environment makes everything very stressful, which ends up taking a toll not only on your life but also on the people around you.

It's the age of technology, and computers are fast replacing man labor. However, the Internet is one of the greatest innovations of this age that man has come to rely on for our day-to-day actions. You can map your way around a city, learn, ask questions, receive answers, make more friends, meet significant people, etc. The benefits are endless and inexhaustible. Be that as it may, an exciting feature of the Internet is one's ability to make money off it.

To understand the law of making money is to know that knowledge is power. That is, what you know and how you apply them go a long way in determining the results you produce in the long run. There are many ways of making money, but have you considered earnings from your computer screen? Rather than let the Internet continually consume you, you can harness its many opportunities and earn legit money.

In addition to its accessibility, there are more than a couple of benefits of making money from the Internet legitimately; it is remotely done, flexible, less time-consuming, and there is no limit to how much you can earn. If you have not been earning online, then you have been letting yourself get cheated of the opportunity to make earnings on a gold platter.

The outbreak of the COVID-19 global pandemic opened many people's eyes to see the importance of making money off the Internet. Due to the restriction on social gatherings, many business owners could not come out to make their usual earnings. However, those people who had had one or two engagements on the Internet before or who took up the advantage that the internet provider experiences an increase in their earnings. People trooped to YouTube, blogs, and social media to keep themselves occupied; they depended on online vendors to provide for their needs. Virtual assistants were needed, and almost everyone worked from home.

Perhaps you have considered making money online to support your paycheck and fund your tuition, but you do not know how to begin or where to begin from. Or you are well accustomed to how the Internet works but still do not know how you can utilize it to make earnings daily, weekly, or monthly. I understand the frustration that comes with this feeling of confusion, and that is why I have prepared this book especially for you, as it has all of your questions mapped out.

In this book, I have broken down the various ways of making money online, giving examples of the different types of business you can choose to venture into, and this includes freelancing, brand influencing, investing, blogging, content and affiliate marketing, owning a YouTube channel, dropshipping, coding, etc. Here I established that you could do anything legitimate online and earn money from it, however minute your pay may be.

This book also waters the ground for making decisions on the best type of business you can go into to earn money. And this revolves around a combination of your skill or skills, technical know-how, passion, and availability of resources. Plunging into a business without first considering all of these factors could be likened to going into a battlefield without gearing up. How well you manage these factors goes a long way to determine how well you can stand adversity and survive amid competition.

Similarly, it is for this reason to stand adversity that I made available steps to equip yourself in preparation for making money on the Internet. You need to take lessons and invest in yourself to improve your old skills and learn a new one if need be.

It is also very important to understand the power of networking and begin to make the proper connection while accessing suitable sites, be solution-driven, develop a money mindset, and access appropriate sites or platforms. Often, people find it challenging to know the right platforms to access when it comes to doing their businesses. They neglect criteria as crucial as knowing if the platform favors their course if legitimate and guarantees their security. For this reason, I gave a list of some of the most popular sites for different means of earning money online.

It is heartbreaking to find people stray right into the hands of fraudulent people on the Internet when they want to make money. You have to understand that not everyone is a fan of making legitimate money, and the fact that you are seeking fair means to earn money does not automatically make you immune to falling prey. You are more vulnerable to these predators. Thus, it would be best if you were on the lookout for the vices of these scammers. This book addresses how to identify them and avoid them all at once.

Perhaps, you already understand your niche and have taken on one or two businesses and means to make money via the Internet; having read this book, you must understand that certain things have been wrong. And I'd advise that you do not quit on what you have been doing, but rather evaluate those strategies you have been using and point out things you think you have been applying wrongly.

If you've correctly understood what this book contains, you will realize that often, the reasons why we fail at reaching our desired goal are because of the limitations we place in our mind and the lack of constraints. It's pretty easy; when we place restrictions on ourselves, in our mind, we do not believe that we can make money online, we do not want to network, we do not trust our skills, and we want to live as it comes. However, for some others, their lack of limitations spurs them into setting unrealistic goals and having misconceptions that you can become a billionaire overnight through online networking. The key often balances, and you have to understand that while you can become a billionaire online, it usually takes time and persistence. Until you know and accept this, you will continue to search for "get rich quick" schemes and fall prey to online scammers.

So, what are you waiting for? Let's jump right into the first chapter now, where I give you a picture of how you can start earning money online in the first place!

Chapter 1

Let's Start With Honesty

How To Make Money Online?

Every day, there are testimonies of people making large sums of money by using their computers. Some people's account balances continue to grow while they sleep, while others have extra earnings on the side as they go about their 9-5 jobs. Perhaps you want to make similar or higher earnings and are at a crossroads as to what type of business to start. This indecision is frequently caused by a lack of diversity in the means, as there are numerous ways to earn money from the comfort of your own home or in your spare time if you so desire. Your means of production could be service- or product-based. In any case, the most important thing is to make full use of the platform.

Blogging

Blogging is one of the most popular means of making earnings online, and there have been testimonies of famous bloggers who hit seven figures solely through this channel. It could be something you naturally enjoy doing or pick up as a skill. And if you already own a blog for your pleasure, you might want to rebrand your blog and make it a money making tool. There are a couple of ways you can earn from your blog. You could use your blog for content marketing and affiliate marketing, or you could make it a channel for advertising some products and get paid while you do this.

Photography

There are many websites and companies out there willing to pay for good photographs. If you are a good photographer, you might want to start monetizing your skills online by advertising pictures of what you do and what you are capable of doing.

Dropshipping

Another legitimate means of making money online is dropshipping. This business is relatively easy and less costly, and it helps you earn some extra cash if you know how to run the business well. Although the pay is usually based on commission, there are testimonies of people making huge benefits from it, and right from the comfort of their homes or side by side with other jobs.

Affiliate marketing

Like dropshipping and content marketing, affiliate marketing provides room for extra earnings online by conjuring with other brands and products for marketing purposes. This way, you get a commission on goods sold. You harness social media platforms to achieve this goal.

Brand influencing

Being a brand influencer can be demanding. Brand influencing is a service-based business that can be lucrative once you get good at it. However, you will have to pay the price of gaining many followers on your social media pages first. However, once you've established yourself, you can make a lot of money online, especially if you become a brand influencer for several company products.

Freelancing

Freelancing is one unique means of making money online. It simply means becoming your boss. There are a lot of services through which you can become a freelancer. You can decide to bring your 9_5 job to the Internet and seek clients that would be willing to pay better for your services. For example, you can render your designing skills as a freelance designer on the Internet.

Content writing

The business of content writing never gets old because now and then, there is always something to write about. Content writing goes beyond advertising, and it is a skill, albeit learnable, that not many people pos-

sess. If you are a good writer, you could consider improving your skills and acquiring content writing skills. With this job, you can work remotely, and you don't have to quit your other job if you had one.

Virtual assistance

There is a minimal difference between a virtual and a physical assistant. There are plenty of organizations, companies, and firms looking to hire virtual assistants who would take up online customer care services, social media management, schedule management, etc.

Graphics designing

One of the most effective ways to make money online is through graphic design. You don't always need to go to an office or meet with your clients in person to get the job done. All necessary arrangements can be made online, and transactions can be completed online. With graphics design, you can have a lot of clients and make a lot of money.

Online tutoring

The power of the Internet cannot be undermined when it comes to learning. Today, there are tutoring sites where students meet with their teachers in online classrooms like Google classrooms and hold actual learning sessions. You can balance this with your regular teaching job if you'd like or take up full-time online tutoring as a job.

Online store

Truth be told, running your store can be tasking. There is a need to rent a place, pay for furnishings and renovation, employ one or two staff, and regularly settle utility bills. However, all of these could be boycotted when you run your store off the streets and online. This is a product-based business with quite a lot of benefits. In addition to boycotting expenses, you can reach a larger audience on the Internet. If you have never done this before, you can start small and see where that gets you.

Social media management

Your social media accounts speak many things about you and your brand. Yet, many companies and celebrities find it hard to manage theirs. The craze for getting a good social media manager is never-ending, and it is also a suitable means of making money online. As a social media manager, you can take up more than one account you handle and get paid without obstructing your regular jobs. You could take the job as a full-time one or a part-time job.

Investment

One way to earn money online from the comfort of your own home is to invest your money or crypto coins. While this is a time-consuming process that will require dedication to complete, it is making money online. Nowadays, there are numerous websites where people can invest their money for future use.

Legitimate online betting

As crazy as it might sound, online sports betting is another way to make money online. The procedure is relatively easy; predict the outcome of a match, place your stakes, and wait for the development bet as development would determine your win.

Content marketing

Content marketing is basically using the same approach of content writing to market products and services. If you are initially a content writer, content marketing skills might come in easy for you to master. People make lots of money becoming professional content marketers. You can become one of such people in your unique way.

Reviewing websites

It might seem a little unreal, but people get paid real money to review websites and test run applications. Reviewing websites does not often require having a unique skill, but you are being paid for your services. The catchy thing about this means is the ability to include it as a side job and add to your schedule their earnings. Additionally, you tend to have fun while getting paid.

Writing codes

Writing codes is one way to make a considerable amount of money over time. Acquiring coding skills often comes with hard work and dedication, but its benefits are countless. Those who know how to code are usually knowledgeable about creating websites and apps. The skill is fast selling among huge companies and organizations.

Online coaching

Like online tutoring, online coaching presents lots of ways to make money online. People spend money on learning, and when you organize a class or two on topics that are bound to solve specific problems, you can earn a lot for your services. The most important thing is finding your target audience and presenting a lecture on a problem that you are sure is eating them up with online coaching. For example, suppose your target audience is fresh graduates from college. In that case, you can organize a class on securing the best jobs and another class on how to compose the best CV to prepare them for the labor market.

There are other ways to make money online, and the list is almost inexhaustible. You can decide to pick one or two related means, work by the tips presented in this book, and begin to earn as much as you can.

The Advantages Of Making Money Online

Infinite possibilities

If you've ever Googled "how to make money online," you've probably come across at least a thousand different options. Aside from the traditional concepts of online selling, copywriting, and participating in paid online surveys, new websites are emerging that will pay you for the strangest things.

Have you ever considered getting paid up to ten dollars for simply visiting a website or sharing something on social media? As the Internet has evolved into a battleground for marketers, so have the benefits for users. You can make a lot of money by reviewing, promoting, or sharing businesses, taking photos around town, or answering simple questions.

And the best part is that it's all accessible via an Internet connection. It necessitates minimal effort and incurs no cost on your part. So, if you have a lot of free time, go check out the options and have fun earning!

Adaptability

There will be no more schedules! You can plan your tasks and working hours around your other scheduled activities rather than the other way around! You can work any day, any hour, and from any location as long as you have an internet connection. You can stop begging for a day off or leaving work early because you are your own boss when you work from home.

This is a huge benefit, especially for mothers who have a full-time job caring for their children and cannot adhere to any schedule. Paid online surveys, which take only 15 minutes and are the least engaging, are ideal for them. Writing content for blogs or freelancing are also viable options for people who have additional responsibilities.

There is no need for investment.

Of course, it depends on how you intend to make money online, but in most cases, all you need is a computer and an internet connection, both of which can be found in any home. You don't need an office; simply designate a working space in your home and transform it into an adorable home office.

Legitimate online gigs will not require any upfront payment so that you can get started right away. However, if you use Gigwalk, you may incur some transportation costs because the app pays you to walk around cities and check signs, labels or take pictures of specific locations.

Completely undemanding

The best thing about making money online is that it requires little effort or skill. Remember when you were a kid, and your parents told you to study hard because people without education will never earn their own money? They'd be sorely mistaken today.

In most cases, you don't need any kind of education or experience to make money online. All you need to put into it is your time. The fact that it takes a lot of time is undeniable, but why not do something productive with your spare time instead of wasting it in front of the TV?

It is both safe and simple.

The majority of online payments are made through PayPal, the most secure and user-friendly online payment system. What's more, you don't have to do anything! The funds are automatically transferred and converted; you do not need to visit banks or cash in checks; they are yours to spend with a single click!

It's worth noting that there may be PayPal alternatives from time to time; just make sure the website is trustworthy and provides a secure connection.

A Statistics Of Online Money Earning

Then there's the possibility of earning significantly more money online. The majority of the best online business models are highly scalable and have low overhead. With you as the sole earner and possibly the sole employee, to begin with, you can generate a massive amount of money and advance much faster than you would by traditionally climbing the corporate ladder.

But isn't it possible that this is all a bit of a pipe dream? Perhaps you do not believe you have what it takes to be a successful internet marketer, website owner, or app developer. Actually, this type of business is more common than you might think. It's also becoming much more common. These online business models are most likely a foretaste of things to come.

We might find that very few of us end up working in stuffy, corporate environments one day. Read on to see some rather impressive statistics about online business to demonstrate just how possible this is today.

Online Earnings Statistics

- 8% of affiliate marketers make $50,000-$100,000 per year.
- Only 12% of affiliate marketers earn more than $100,000.
- Many affiliate products pay up to 75% commission!
- Over one-third of businesses are implementing a "digital transformation program" to capitalize on the promise of making money online.
- Another third has already implemented such a program!
- In 2015, content marketing produced three times the number of leads for businesses as traditional marketing.
- Video marketing is used by 73% of B2B organizations.
- 78% of businesses have dedicated social media departments.

- 91 percent of all consumers check their email daily.

- Sixty-six percent of US customers over the age of 15 have purchased a product due to an email ad campaign.

- Email marketing has a 4,300% return on investment!

- One out of every five Americans now works from home (thanks to the Internet)

- This equates to 53 million self-employed Americans.

- 54 percent of marketers believe SEO will become more popular.

- Search engine optimization is used by 96% of marketers.

Surprisingly, search engine optimization is the most popular marketing-related search term on Google.

During the first half of 2017, search ads accounted for $9.1 billion in advertising spend.

Disadvantages Of Online Money Making

Isolation in part

When you work for yourself and make money online, you most likely work from home or a secluded location to be alone and complete your tasks. This can result in a lack of social contact as well as feelings of loneliness and isolation.

Furthermore, since we already spend so much time in front of computers, imagine making your living in front of them as well. This is undeniably harmful to your eyes, sleep, and overall health.

Be wary of con artists.

Needless to say, for every great online earning opportunity, there are at least twice as many fake or scam ones. You must be extremely cautious when screening them out so that you do not end up losing more money than you earned.

- Scammers are easy to spot if you follow these simple guidelines:

- Never agree to pay anything; it's a job, and they should be paying you.

- There is no such thing as a free meal; do not fall for advertisements that appear too good to be true.

Always conduct a background check on the company you're considering.

What To Remember Before Moving Forward

In the world today, there is nobody perfect, and we make mistakes when it comes to different areas of our lives, in our relationship, car-

Rushing into an online business

Rushing into an online business is one of the most common mistakes for beginners who aim to optimize internet use. It is okay to have a zeal for making money and, in fact, begin to make money as soon as possible. However, it is not okay to dive headfirst into a business without adequately researching and gathering vital information for the business you have placed your focus on. In the preceding chapters of this book, I identified the tips to take before deciding what business to begin and how to prepare yourself to make money. And unless you apply those principles, your business might end up hitting rock bottom.

Falling into the hands of online scammers

It is a painful ordeal to fall into the hands of online scammers. It means a waste of time and resources, and it could hurt your mental health, mainly if you were way too deep and finding your way out of the mess becomes too difficult. Often, we become victims of these predators when we rush into businesses, making unrealistic goals to see ourselves in the nearest future. Online scammers usually have no pity whatsoever and do not care if you are a beginner in the business or just an ignorant fellow. What matters to these people is that they get what they want. To them, it is a job, like you are out job hunting. The tendency of falling into the hands of online scammers is another reason why you should be careful of their schemes. I will be teaching you how to uncover their schemes as we proceed.

Having misconceptions that making earnings online is easy.

One of the many misconceptions people have about making money online is that it is an easy thing to do. For some, they believe that it should be way easier than making money through physical means. This misconception does not help you in your journey to success, but rather, it plants in your impatience. It is often said that nothing good comes easy, and like every other business, you will have to work hard to get your earnings. Let your thinking be channeled towards putting your best efforts into whatever means of making online earnings that you choose to adopt.

Making unrealistic goals

If I ran a survey asking people who venture into one form of online business or the other to make money, what they plan to achieve from their businesses, I'd get appalling responses. Many people go into online businesses with the primary objective of becoming millionaires and even billionaires overnight. For example, an investor wants to put in a dollar on Forex and receive a thousand dollars in a week. It doesn't sound right, and it is for this reason that many people fall into the hands of online scammers. Let it be known that making money off the Internet is not a day's job, and it would require some time for your meal to be well cooked and ready to be served.

Boycotting the normal process

Trying to boycott standard procedures to the top will only expose you to losses and online scammers. It would be best if you were careful how you thread in whatever business you choose. For example, you might be required to send applications to your clients before getting a significant deal. Trying to override this process might result in you wasting your time and resources in your bid to skip this phase.

Indulging in too many activities at the same time

We frequently interpret the ability to handle more than a couple of affairs concurrently as proof that we are skilled at multitasking. Being diverse is not a bad thing, but it becomes a problem when you cannot properly manage your business. Many people believe that the more business they engage in, the more money they will make, only to discover that they are not making nearly enough money due to a lack of proper management. When you have too much on your plate, you pay less attention to everything on that plate when you distribute your attention evenly. Others make the mistake of taking on businesses in which they have no experience. It is only fair to you to be involved in what you are capable of handling.

Making Money Online Scams

It is not unusual to encounter one or two people who try to defraud you as you search for a viable means to making earnings online. Sometimes, you might fall victim more than once or twice. Online scammers are sometimes tactical and crafty at deceiving their victims. While some of them have visible pointers to earn you to steer clear, you might need to be smart to uncover others. Be that as it may, for some people, even if they have glaring warning signs on some sites, they would still fall victim to their vices. Falling victim to these scams often happens not because people like to be scammed but rather because they are ignorant of these pointers. Here are some tips to help you point out scammers when you encounter them.

No contact information or license

As a money-minded person, you need to be intelligent and vigilant to not fall into the hands of swindlers. The first red light warning sign to look out for is the contact information. Before you trust a site enough to carry out one or two businesses with them, ensure an actively working contact comprises an email and or a phone number you can reach. Usually, there is a licensed number that proves they are certified. You might like to go through the process of verifying that they are accurate and not fake.

Site redirect

When you click on a link from a site and you are redirected to other irrelevant stuff, you should know that there is something wrong somewhere. Sometimes, this problem might be due to technical failure and not necessarily because the sites are fake, but you do not want to flag off this warning sign. Take the extra effort to do some research online before continuing the process. Trust me; it would be worth it as you never know what you might find.

The "make money fast" scheme.

One of the vices used by scammers to entice you into their trap as a beginner in online businesses or even an old-time visitor is the tactic of making quick money as soon as possible. Although there are some ways to make money while traveling, there is usually no way to double your money and have your earnings the moment you invest it. Everything has procedures, and legitimate businesses take time. For example, suppose you send in a completed job as a content marketer. In that case, your clients must approve your services before paying in full.

Auto-renewal

Auto-renewals are often normal procedures and suggestions when trying out online services. They are used in apps, subscriptions for online goods, etc. However, some shady websites are super quick at requesting your card details and asking that you renew a subscription, even when you do not get the value of what you are paying for. Again, you must understand every phrase and clause before you click that button.

The "Pay for one trial and get your money back" guarantee.

Most times, when you come across ads like this, it is almost certain you're walking into the net of scammers and that you will not like the trial, and also, you will not be getting back your money. It is important to note that all of these schemes could have a legit backing to them, but it all boils down to your willingness to research and find out what needs to be found.

Fake comments and reviews

It is not new to find sites that go through hiring people to give wrong reviews. Also, there are cases of photoshopped comments and testimonies about how well the site has helped their online business grow and how much money they have been able to make in a short unrealistic period. This tactic often works out fine, and you find many people signing in to register under them. For example, you make content to help market a product that probably doesn't exist or is of lesser quality. Or in some cases, you do not receive payment for services rendered, and you find your work being used without your consent.

Guaranteed instant winnings with no losses.

It is never guaranteed that you will be on a winning streak with no losses, no matter how good you are at what you do. There will be times when you do not make sales if you are in the product business. Similarly, your client may reject one or two of your jobs if you work in the service industry. And, as an investor, the market may not always be on your side. It does not always imply that you cannot continue to be favored, but it does imply that you cannot always predict the outcome of your business. However, if you enjoy taking risks, you should exercise caution in how you do so. When you sign up for an online business, you may be asked to pay a token to help you get started.

For example, becoming a brand marketer may entail purchasing a starter pack for a certain amount of money. However, this system is one of many ways to be duped online. And instead of making more money, you end up losing what little you have. Before filling out any forms or disclosing sensitive information, you should carefully consider what you are about to do.

It is best to choose more popular sites because there are fewer chances of being scammed when it comes to their reputation. Making money online isn't always a bed of roses, as there are ups and downs, which is why you need to know where you fit in. When things go wrong, you need someone, a body, to whom you can lodge complaints and receive feedback.

For example, as an online freelancer, you successfully completed one or more jobs for a client on time. However, your job is rejected for reasons unknown to your client, and your fee is withheld. In the future, you will probably discover copies of your one-of-a-kind finished original work in total circulation with no credit ascribed to you. At this point, you have the right to file complaints and have them resolved through a registered and licensed platform.

When we sign up for a platform to make money, we often trust that our money and information will be safe and secure. This is why you should not hand over your personal and confidential information to con artists on a silver platter.

Chapter 2

Affiliate Marketing – A Single Most Profitable Method

What is Affiliate Marketing?

Simply put, affiliate marketing is a system or, more accurately, a type of marketing in which you, as an affiliate marketer, earn a commission by promoting products from other people and businesses. These commissions are paid out each time a customer purchases a product or service based on their recommendation.

There are several ways to earn a commission, but the most common is to place affiliate links on your website or blog. You earn a percentage of the total price of a product when someone purchases it through that link. The commission rate you receive is determined by the type of product you promote. For example, suppose you promote PC and electronic devices on Amazon's affiliate program. In that case, you will earn 2.5 percent for each customer you bring through your affiliate link. On the other hand, promoting women's fashion and jewelry can earn you up to 10% of every sale.

Because you'll want to track those conversions, each affiliate link is unique and has its own code. Otherwise, it would be impossible to determine which associate referred the customer. What does this have to do with you? This means you have the opportunity to profit by selling a product that is not your own. The more you sell, the more money you make. It's a fantastic deal because you don't have to worry about shipping, overhead, or customer service!

To make money as an affiliate marketer, you must first understand all parties involved and hope to gain from the relationship. Knowing each role and how it can potentially help you make money is an important first step when getting started.

How Does Affiliate Marketing Work?

Overall success necessitates the development of relationships with three distinct parties: Advertiser, Publisher, and Consumer.

Advertisement

This is the party that is selling the product or service. An advertiser is a person who created a product and is responsible for selling it. It could be anyone, from a solo preneur to a large corporation. This is the party with whom you will be collaborating as an affiliate marketer.

Affiliate

Affiliate This is the person who works with the advertisers/merchants to sell products in exchange for a commission. The relationship between the advertiser and the affiliate is critical. Because both parties need to earn money for the relationship to continue.

Consumer

This is the party that will (hopefully) purchase the product you promote. As a result, the affiliate-consumer relationship should be based on trust. Will the customer be aware that he is purchasing from an affiliate marketer? This is entirely dependent on the marketer.

Thus, the marketer decides whether or not to be transparent and mentions that he or she promotes a product through an affiliate program. In some countries, this may be required by law. In either case, the customer is unaffected in any way. Customers are most affected by reliance and affiliate credit.

Network

The network acts as a go-between for advertisers and affiliate marketers. Typically, it allows advertisers to sell their products and affiliates to select the desired affiliate program. As previously stated, Amazon Associates is one of the most well-known networks of this type. Every party benefits when all aspects of affiliate marketing work together. Consumers receive their product, merchants earn money, and the affiliate earns a commission.

How Does One Get Paid?

Affiliate marketing is an undeniable draw for those looking to increase their income online because it is a quick and inexpensive way to make money without the hassle of actually selling a product. But how does an affiliate get paid after introducing a seller to a customer?

The solution can become complicated.

The consumer does not always have to purchase the product for the affiliate to receive a commission. The affiliate's contribution to the seller's sales will be measured differently depending on the program.

The affiliate may be compensated in a variety of ways:

Pay Per Sale

This is the most common affiliate marketing structure. The merchant pays the affiliate a percentage of the product's sale price after a consumer purchases the product due to the affiliate's marketing strategies. In other words, the affiliate must actually convince the investor to buy the product before they are paid.

Pay Per Click

Pay-per-lead affiliate programs are a more complex system that compensates the affiliate based on the conversion of leads. The affiliate must persuade the consumer to visit the merchant's website and perform the desired action, filling out a contact form, signing up for a product trial, subscribing to a newsletter, or downloading software or files.

Pay Per Click

This program encourages affiliates to direct customers from their marketing platform to the merchant's website. This means that the affiliate must engage the consumer to the point where they will leave the affiliate's site and visit the merchant's site. The affiliate is compensated based on increased web traffic.

Learn About The Different Types Of Affiliate Marketing

Do you want to start an affiliate or partner marketing program? Excellent! Are you perplexed about the different types of affiliates you can find and work with?

You're not by yourself. There is a lot of information available. It can be difficult to determine which type of affiliate is best for your company. You may discover that your specific affiliate or partner program is only applicable to a small number of partners. Alternatively, your program may be adaptable enough to affect them all. Knowing your target market is always essential for a successful business.

Examine the affiliates who have signed up for your program to see if you can spot any patterns. Do those websites have a lot in common in terms of style, or are they all over the place? Depending on your target audience (or potential target audience), you may want to create several distinct campaigns to promote your company. Pursue new types of affiliates or streamline your efforts to focus solely on one or two of the most powerful types.

But, exactly, what are the various types of partners or affiliates? Thank you for inquiring! We've compiled the following list to give you a good overview of the various types of affiliates spreading the word on the Internet. Keep in mind that some sites may fall into more than one category.

Coupon Sites

Coupon or deal sites are without a doubt the most popular type of affiliate site available today. And for a good reason: they are eCommerce-friendly websites that consumers visit when looking for a good deal–which is the majority of online consumers. Affiliates will provide coupon or discount codes to their users in exchange for commissions on any sales generated by someone clicking through one of their links to a merchant site. RetailMeNot, Offers.com, Coupons.com, and Red Plum (soon to be RetailMeNot every day) are excellent examples. Some even allow users to upload their own coupon codes.

News And Media Sites

Yes, news and media websites are becoming affiliates. While it does not help "fake news," many large media companies are monetizing content and using affiliate links to do native advertising within articles. Media conglomerates such as Ziff Davis are regularly acquiring online publications, while others are diversifying revenue streams to combat declining ad sales. In any case, these are major players in the game.

Content Sites

Gossip, entertainment, sports, and lifestyle, among other things. When a website is compared to a magazine, you know it has a lot of content. These sites regularly publish content articles and columns, monetizing with affiliate banners and links and, at times, sponsored content. Entirely user-generated sites, such as discussion forums, can fall into this category because the entire site is made up of user-generated content.

Review Sites

These websites evaluate various advertisers, usually in a specific niche. One of the most popular places to find these are on Amazon, and affiliates and partners are told that setting up an Amazon review site is one of the best ways to grow their business and start making money. But don't just pick anyone. Look for keyword-rich URLs, clean, uncluttered content without many popups or widgets, and well-thought-out reviews that are tailored to the reader's needs.

Loyalty/Cashback Sites

Cashback sites are also very popular among users. They actually refund money to members based on purchases made via the site's affiliate links. Essentially, the cashback site received a commission on every sale, but now they return a portion of their commission to the user who made the actual purchase. It results in a lower profit for the publisher with each sale, but they more than make up for it by heavily encouraging members to buy through various promotions and incentives. eBates is the most well-known and widely used.

Blogging

Bloggers are a subset of content types, but many blogs specialize in particular niches. These range from personal interests to consumer product reviews to advice on topics as diverse as saving money and parenting. Depending on the blogger, affiliate banners or affiliate links may be included in their blog posts. But keep in mind that not all bloggers are created equal. Choose bloggers who discuss issues relevant to your products with your target audience.

Shopping Comparison

Sites that provide eCommerce tools, such as comparison shopping rely heavily on data feeds containing products linked to affiliate links. Users can search for millions of products on sites like Shopping.com, which returns relevant results. When a user clicks through an affiliate link to purchase a product on a merchant's page, the comparison shopping site earns a commission.

Email

Despite the popularity of text messaging, people still prefer to communicate via email. And many affiliates profit from it. Large publishers have built email lists to reduce their reliance on Google traffic and the penalties associated with algorithm updates. Email campaigns aimed at a specific customer demographic are common, including multi-merchant email opportunities featuring deals or special promotions. However, Caveat venditor (Latin for "let the buyer beware"): Working with email affiliates can be dangerous, especially if they aren't CAN-SPAM compliant. Affiliate fraudsters will sometimes classify themselves in networks as "Email" affiliates because their promotional activities are less visible to the merchant/advertiser.

Mobile

Certain websites incorporate affiliate links into their apps. This applies not only to coupon or shopping comparison sites that have an app but also to standalone apps. Popular apps such as Shazam and Pandora use affiliate links to direct users to Apple's iTunes Store, whereas free gaming apps monetize through incentivized offers.

Search Affiliates

These are business owners who use search engines, Facebook, and other paid advertising sites.

The Steps To Starting Affiliate Marketing

To become an affiliate marketer, you only need to become a good affiliate marketing program member and start spreading the affiliate link. To make money consistently, however, you need to be smart in your approach to the business.

Here are the things you will need to start with this business:

A website

As stated in the previous chapters, most affiliate programs require a website before you will be accepted. Advertisers check the website to see what types of content you have. Some advertisers are extremely picky when choosing affiliates. They only accept those whose websites can bring in huge amounts of traffic. Don't worry, though, because beginner-level affiliate marketing programs will accept even publishers with new websites.

A profitable niche

You cannot just create any type of website if you want to become a successful affiliate marketer. In particular, you need to be careful in choosing the niche market. We will discuss how you can choose the right niche market in future chapters. For now, just remember that the types of affiliate marketing products that you can sell will depend on the niche market you choose to participate in. Suppose you want to sell only high ticket items, for instance. In that case, you need to make sure that the niche market you've chosen has high-ticket affiliate marketing programs.

An organic source of traffic

In the world of internet marketing, there are two ways for you to get traffic, organic and paid. While some affiliate marketers use paid methods to earn through affiliate marketing, these methods expose you to a higher financial risk. Those who use these methods have developed their skill over years of trial and error.

For now, you should focus on building assets that will help you gather organic visitors. This includes social media accounts and pages, forum memberships, accounts in niche-specific online communities, offline traffic sources, and other similar assets. You will need to think about where you will get your traffic for your website if you want to be successful from day one.

A membership to an affiliate marketing program

There are tons of affiliate marketing programs on the web. However, not all of them will be suitable to offer the type of traffic that you can get. You will need to choose an affiliate program that fits the needs of the audience. We will discuss how to choose the best affiliate marketing programs for your audience in the following chapters.

The four factors above are the minimum requirements for becoming a successful affiliate marketer. After choosing an affiliate marketing program, you can apply directly from the web. Most of the application processes just require you to fill up a form with your personal details. In addition, the application form may also ask how you plan to generate traffic for your affiliate offers and what types of products you wish to promote. The form may also ask you how much traffic your own website gets a day and what other monetization methods you use in the said website.

In addition, you may be asked to fill up a tax form. This part will depend on the requirements of the country of origin of the affiliate program. When signing up for Amazon Associates for Amazon.com, US citizens will be asked to fill up a US tax form. A different form will be required if you are applying for Amazon.ca (Canada) Associates program.

How hard is the application process?

For some affiliate programs, getting in is easy. Some may even automatically approve your application. There are more specialized programs, though, that will ask for more requirements after you apply. The most demanding programs will have their employees scan your website for the quality of the content. If they are not satisfied with the design or the quality of the content of your website, they may reject your application.

These extremely selective programs usually do this for different reasons. The majority of them do it because they want to protect their brand. Big brands only want to be associated with websites that have good quality content. Some of them will only accept websites that have been around for a long time with hundreds of archived content. Some brands will also reject your application if they see that your website or content is not aligned with their target consumers. Suppose your website is in English, for example, and you apply with an affiliate program for Spanish people. In that case, you are likely to be rejected.

My application has been approved; what now?

It is after the application process that the real work begins. Now that you have a product or service to promote, you can start gathering traffic and funneling them towards your affiliate links. There are multiple methods on how you can do this.

First, you can use paid methods as a source of traffic. You can use ad-advertising platforms that allow affiliate links and pay for the clicks or the views that your ad gets. With this method, you are spending money to make money. You will need a bankroll of hundreds, if not thousands of dollars, to make this strategy work. You will also need to make sure that you follow the guidelines set by the affiliate advertisers.

Most of them will restrict you from using certain keywords in advertising targeting. If you are selling Nike rubber shoes, the affiliate program contract may restrict you from using the keywords like Nike. They do this to prevent affiliate marketers from competing with the mother company in the advertising bidding.

The second method of sending traffic is to spread the link around the web. In the past, people used to do sleazy techniques to get clicks on their affiliate links. Some of them, for instance, put their affiliate links in the signature part of their forum accounts. This way, when people see their forum comments, they also see the affiliate links. Many forums learned about this technique and now ban the use of links in the signature.

Some affiliate programs also allow the use of redirects and pop-ups to gather traffic. An affiliate marketer using redirect may set up a page that will automatically redirect to the affiliate website. When the visitor visits that page, they are automatically transferred to the affiliate program's landing page. The problem with this strategy is that most people who go through the automatic redirect are sent to the affiliate program's website unwillingly. The majority of them will bounce (i.e., leave rather than see other pages on the website). This excess traffic of non-buyers will eventually take its toll on the advertiser's website. This is the reason why many affiliate programs ban the use of redirects. Some people also use link pop-ups to send people through the affiliate link. They may set the pop-up to appear after a certain link in the website is clicked. Upon clicking the said link, another window or tab will open. This works both for desktop and mobile browsers. When the pop-up opens, the affiliate link is triggered, and the affiliate landing page starts to load.

This process of sending traffic to affiliate programs is also problematic. Aside from being prohibited by most affiliate programs, browsers also tend to have pop-up blocking technology. This prevents most of the pop-ups from opening even when the right link is clicked.

The recommended way of gathering traffic for your affiliate programs is through the use of content marketing. Simply put, content marketing is the process of gathering internet user's attention by using different types of content media. One has the option of using text, audio, images, videos, or a combination of all these to invite people to go to your website.

News websites are a classic example of websites that use content marketing. They create news articles and accompany them with videos and relevant images. After creating them, they post their content on their social media properties. They spread their news articles through

Facebook, Twitter, and other social media marketing platforms. When a person interested in the news sees the article, the headline and the accompanying image should compel them to click on the link of the article. This will lead them to the news website.

After reading the article, they may press the up-vote button for the content. Some may even share a link to the content with their friends. By doing this, they are spreading the news and increasing the reach of the news article. Other people interested in the news may also click on the article. Afterward, they may also share the content with their own social media followers, continuing to spread.

Some people looking to read that specific news article may also go to Google and do a search. Because of proper search engine optimization practices, the article landed on the top spot page with top spot relevant keywords or key phrases. Because of this, more people from Google manage to read the content. Some of the readers interested in the content of the article may also choose to share it with their friends.

While this may seem like a simplistic illustration of how content marketing works, this is how it happens for most content on the web; content creators simply make the content and share them in the relevant online hotspots.

They develop their skill in getting people's attention over thousands of hours of practice.

You could also do the same with your affiliate marketing business. You could lure internet users to go to your website by creating and sharing content relevant to their interests and needs. In viewing your content, they should also see the affiliate ads that you share on your website. A percentage of them will click on these marketing materials and make a purchase. You will receive a percentage of the sale amount that your referred customers spend.

The Basic Affiliate Marketing Terminologies

API

API is an abbreviation for an application programming interface. The API is a collection of functions and procedures used by affiliates and advertisers who want to obtain automated data from Adtraction's system rather than manually signing in and obtaining reports. Their API is "conversing" with our API.

CPA

Cost per action and cost per acquisition are two terms that may be used interchangeably. CPA is a payment model in which affiliates are compensated for performing a specific action. An action could be to generate a lead or a sale, for example.

CPC

CPC (cost per click) is another payment model in which advertisers pay their affiliates per click received.

CPL

The cost per lead (CPL) is closely related to the cost per action (CPA), which refers to the specific action of generating a lead.

CPM

Cost per mille is an abbreviation for cost per thousand impressions. In this case, the advertiser pays a set price per 1,000 activities.

CPO

The traditional payment model for affiliate marketing is CPO (cost per order). Affiliate marketing is a performance-based marketing model in which the advertiser pays a set commission per order to which an affiliate contributes.

CR

The conversion rate is the percentage of clicks that result in the desired action out of the total number of clicks. If your tracking link received 10,000 clicks and you made 100 sales, your conversion rate is 1%.

CTR

The acronym CTR is an abbreviation for The click-through rate indicates how frequently people who see your advertisement click on it. If you have 1,000 website visitors to your blog and 100 of them click on your ad, your CTR is 10%.

EPC

EPC equals earnings multiplied by the number of unique clicks.

Google Analytics

Google Analytics is a web analytics service.

Google Analytics is a free web analytics tool that generates detailed statistics about the visitors to your website.

Google Trends

Google Trends is a search engine that analyzes trends.

Google Trends is a trending search feature provided by Google. The feature displays the frequency with which a specific term is entered into Google's search engine over a given time period.

Google Tag Manager

Google Tag Manager is a web-based application that allows you Google Tag Manager is a tool that allows you to manage and deploy code snippets or tracking pixels on your website without the need for code modification.

Lead

A contact who has the potential to become a customer.

LR

The LR Lead rate is the percentage of leads generated based on the total number of unique clicks. If 1.000 unique clicks resulted in 10 leads, your LR is 1%.

Leads multiplied by the number of unique clicks equals LR.

Pixel

The term "pixel" refers to A tracking pixel is a piece of code that allows for the tracking of website visitor behavior. A pixel is loaded when a user visits a website. Advertisers can use tracking pixels to collect data for online marketing, web analytics, or email marketing.

PPC

PPC stands for Pay Per Click.

PPC stands for pay-per-click and refers to a model in which advertisers pay a fee for each click on one of their ads. It is closely related to the CPC.

SEM

SEM is an abbreviation for Search Engine Marketing.

Search engine marketing can encompass search engine optimization, paid advertisements, and other search engine-related services.

SEO

SEO is an abbreviation for search engine optimization. SEO is the process of increasing the volume and quality of traffic to your website via organic search engine results.

Tracking Link

A tracking link, also known as an affiliate link, is a unique URL to which an affiliate directs an audience for a specific offer.

The Advantages Of Affiliate Marketing

According to Statistica, affiliate marketing spending in the United States alone is expected to reach $8.2 billion by 2022.

Why do businesses spend so much money on affiliate marketing campaigns? Some of the advantages of using affiliate marketers in your business are as follows:

Low startup costs

An affiliate program does not necessitate the creation of ad visuals or the purchase of ad space.

Instead, you'll have to rely on your affiliates to create marketing content for you. Other than the initial effort of selecting and vetting affiliates, marketing your products requires little effort from you, which is one reason it has become such a popular marketing method.

Once a company is comfortable working with an affiliate and has established a good relationship, they can largely delegate your service or product to them.

Low on-going expenses

Because affiliates perform most of your marketing activities, they will also bear a portion of the cost. And, because it's a commission-based program, you'll only pay affiliates based on the sales they generate.

Unlike other marketing models, such as PPC advertising, which requires you to pay for every click that leads to your website, this marketing model is easy to manage and does not disrupt your cash flow. On the other hand, affiliate marketing ensures that you only part with your money when sales are made.

Low danger

Low risk equates to a low cost. Payments are only made when there is an actual conversion, so there is little risk of loss, ideal for businesses on a tight budget.

Focused Traffic

Because you handpick your affiliates, you can ensure that the traffic that comes to your site is from people who find your product or service useful. This is because affiliates who resonate with your brand will almost certainly have people in their sphere of influence who will find your brand useful.

Adaptability

You can easily scale up or down your affiliate program for little or no cost. It also provides you with an excellent opportunity to expand your business without breaking the bank.

A high return on investment

Having the keys to product innovation and effective management skills isn't enough to ensure a high ROI; you'll also need to nail your marketing.

Affiliate marketing has been shown to have a much higher ROI than the majority of marketing strategies.

One important reason for this is that your target audience will learn about your product or service from affiliates in their spheres of influence. In short, you're channeling your marketing efforts through an affiliate who has direct access to and influence over a highly targeted audience that's interested in what you're offering.

How Much Money Can You Make Through Affiliate Marketing?

You can earn a decent living from affiliate marketing if you consistently drive traffic to your website and affiliate links. One significant advantage is that once you rank in search engine results and attract organic traffic, you can continue to earn money for months, if not years, after you first publish your content. While some people supplement their existing income, many affiliate marketers go full-time after building their website and generating significant streams of affiliate marketing passive income.

But, how much do affiliate marketers earn?

VigLink conducted an in-depth survey in 2017 to determine how much affiliate marketers earned in 2016. According to the findings, 9 percent of affiliate marketers earned more than $50,000, and 65 percent earned 5 to 20% of their total yearly income from affiliate marketing.

According to the survey, the longer you have been affiliate marketing, the more money you are likely to earn per year - 60% of the top-earning affiliate marketers have been running their affiliate business for five years or more.

The top affiliate marketers earn more than $2 million per year. Jason Stone generated $7 million in retail sales through affiliate marketing in a year.

Chapter 3

Build Your Own eCommerce Website

What Is eCommerce?

Ecommerce, also known as electronic commerce or internet commerce, refers to the purchase and sale of goods or services over the Internet and the transfer of money and data to complete these transactions. Ecommerce is frequently used to refer to the online sale of physical products, but it can also refer to any type of commercial transaction facilitated by the Internet.

Whereas e-business encompasses all aspects of running an online business, eCommerce focuses solely on the exchange of goods and services.

On August 11, 1994, the first online sale occurred when a man sold a CD by the band Sting to a friend via his website NetMarket, an American retail platform. This is the first instance of a consumer purchasing a product from a business via the World Wide Web—or "ecommerce," as it is now commonly known.

Since then, eCommerce has evolved to make it easier to find and buy products through online retailers and marketplaces. Independent freelancers, small businesses, and large corporations have all benefited from eCommerce, which allows them to sell their goods and services on a scale that traditional offline retail does not allow.

By 2020, global retail eCommerce sales are expected to reach $27 trillion.

What Are The Benefits Of Building Your Own eCommerce Website?

eCommerce is expected to grow from $2.3 trillion in 2017 to $4.88 trillion by 2021, with the majority of that coming from popular marketplaces, but there is no shortage of opportunity for your own owned online store.

At the very least, having your own website diversifies your online presence. It protects your brand from being at the mercy of a single marketplace and its ever-changing rules. A website is, at best, an owned channel that enables eCommerce sellers to increase profit margins and conversions.

Your store can gain trust on large marketplaces with large customer bases, but once you've reached a certain level of success, it's time to build your own website.

Migrating to other platforms, including your own, is a natural and necessary step in the growth process. It allows you to take control of your brand and protect it from platform rules and suspension, even as those platforms evolve.

Here are a few reasons why every successful online retailer should invest in their own website.

Having ownership of your brand entails having ownership of the customer.

Moving away from selling as a third-party seller on Amazon or eBay means regaining control of your brand as it expands to new platforms. When you establish that brand on your own website, you tell customers that they are purchasing from you rather than an eCommerce platform. Because you own the sale and thus the customer, you have more opportunities to drive brand loyalty.

When you sell something on your website, the customer is buying directly from you. They are familiar with your website. When they buy from your store on Amazon, eBay, or Walmart, you can brand your products, drive customers to your original website, and maintain control as the brand behind the sale.

To drive customers to your website, brand your products and boxes wherever it does not conflict with the Terms of Service.

When it does not conflict with the platform's TOS, add your website URL to your pages and logo wherever possible.

Connect with customers as a brand, not as an individual.

Multi-channel fulfillment does not have to be difficult. Delivery provides end-to-end fulfillment, transparent pricing, and quick delivery.

Branding your store allows you to develop a customer base of people who know and trust you, who expect certain service levels or quality, and who will recommend you to their friends and family. Because you have a concrete, well-defined, and focused online presence on directing customers to, having your own website makes branding easier and more reliable.

Diversification: Provide an Omni-Channel experience while minimizing risk.

If your entire brand is invested in a single platform, that brand could be lost at any time. For example, Amazon occasionally suspends even healthy accounts with no red flags. While those accounts frequently reappear, the business still loses money and customers while they are down. Brands must transition from a single channel to an omnichannel model.

Moving your eCommerce business to multiple channels helps you spread risk so that if something goes wrong, your business is still protected, and you are still making money. Furthermore, if your existing marketplace customers recognize your brand, they can find your website and directly purchase from you. Having your own store provides customers with multiple ways to contact your company.

Promote brand recognition and loyalty.

Repeat sales are one of the primary drivers of eCommerce growth because selling to an existing customer is easier and approximately 20% less expensive than pursuing a new business. Because you can establish your own eCommerce store as the source of the customer's satisfaction and value, selling from your own website provides you with more opportunities to create new customer touchpoints and repeat business.

You can do the following with a web store:

- Sign customers up for email updates so they can be contacted again in the future.

- Drive customers to your social media pages to keep them engaged.

- Connect and provide value through product-related content such as videos and blogs.

- Market a branded line of products that, ideally, will inspire the same consumer trust as your original product, resulting in repeat sales.

- Use package inserts and coupons to encourage customers to return or order accessories/refills.

Because studies show that repeat customers account for approximately 65 percent of all brand sales, each of these tactics can significantly boost your ability to drive sales. On the other hand, customer contact is frequently not permitted on platforms such as Amazon, where you are not permitted to use branded package inserts or send emails beyond what is required to provide a quality customer experience.

Expanded marketing possibilities

Revenue can be increased through cross-promotional marketing powered by your own website.

For example, suppose you sell coffee filters. In that case, you could collaborate with a coffee machine brand to offer their customers a 25% discount on their first purchase of filters. This would increase the value of your brand, your partners, and your customers.

This promotional marketing opportunity will not work if you only use one platform, such as Amazon, because you must rely on the platform's capabilities, policies, and the customer's willingness to make another purchase through that platform.

Lower your cost per sale with your own store.

Any marketplace where you sell will charge you a percentage for using their platform. While creating and managing your own online store is costly, you may quickly discover that it reduces costs or pays for itself as you drive more sales through it.

Suppose you can reach a customer directly through your own website. In that case, you will save on seller marketplace fees and be able to lower your overall average cost per sale.

Better search marketing

Owning a website enables you to use a variety of marketing methods that a platform does not allow.

SEO and SEM, for example, can both drive traffic and customers to your store. You can optimize for search in a search engine on your own website rather than optimize for search within a marketplace. On a website with millions of other listings, optimizing marketplace listings for search will only get you so far.

Launching your own eCommerce store also allows you to use blogs and videos to drive traffic, generate leads, and strengthen relationships with current customers. All of these activities have the potential to generate massive revenue over time.

Begin private labeling.

Private labeling is becoming increasingly popular among eCommerce sellers. If you own your own brand and channels, you can sell whatever you want. Even if you start with generic products, branding and selling them across your platform and sales channels will provide you with more opportunities to build trust and repeat business.

More importantly, as a private label, you will have more opportunities to collect customer data, allowing you to develop better products over time and provide ample opportunities to customize services/products for your customers.

Gain the ability to sell wherever your customers are.

Any eCommerce platform will have limitations, restrictions, and frequently difficult requirements that will require time and money to meet. Your own store provides you with more options for selling how and what you want.

A website also allows you to sell anywhere you want. Pop-up stores or brick-and-mortar stores, such as Anker and Bonobos, can help drive sales. You can offer customers the option of picking up their orders in-store. You can actively engage with customers in ways that eCommerce platforms would never allow or support.

Owning your own store enables you to investigate independent logistics options while maintaining consistency in shipping wherever you want. Shipping quality and coverage are important factors in determining the success of any eCommerce seller in a competitive market.

Even if they are only resellers, every eCommerce seller should have their own website. You will be able to drive additional sales, improve branding, and lower cost per sale with the right strategy in place and the potential for private label products in the future.

Most importantly, profit from your website will not be realized solely through sales. Your website will increase consumer trust and awareness, encourage loyalty and retention, assist buyers in making better decisions, and work to push sales across all of your existing platforms.

The Different Types Of eCommerce Websites

With the current trend of consumers purchasing online and via mobile devices, there is no better time to launch an eCommerce business.

While you may have good intentions of getting your online store up and running, there is a learning curve involved. And, as a starting point for your research, look into the various types of eCommerce business models and websites.

It will be easier to decide how you will run your online store and monetize your business once you know what business models are available. Let's take a closer look at the various eCommerce business models.

Business To Business

The structure of the business-to-business eCommerce model is exactly as it sounds. It is the location where businesses sell their wares to other businesses. Products can be anything that improves the business practices of another company.

A software company is an excellent example of a B2B business. Consider Salesforce, HubSpot, and SurveyMonkey. These products are for the benefit of businesses, and the average mom-and-pop consumer would not make sense to invest in enterprise software.

Business To Consumer

The business-to-consumer eCommerce model is simple as well. It is when a company sells products directly to consumers online or through a mobile device.

Business-to-consumer eCommerce companies are those that you and I visit every day to buy things like clothing, books, makeup, and so on.

Amazon, eBay, and Alibaba are among the top B2C eCommerce companies.

It's also important to remember that B2C businesses don't have to exist solely online, as Amazon and eBay do. B2C businesses can also open a physical store and, if appropriate, launch an eCommerce website. Large fashion retailers such as Nordstrom, J. Crew, H&M, and others are examples of this.

When one consumer creates and sells a product to another consumer, this is a consumer-to-consumer business model.

A consumer may sell products to another consumer through a platform such as Amazon, Etsy, or eBay.

Consumer To Consumer

When one consumer creates and sells a product to another consumer, this is a consumer-to-consumer business model.

A consumer may sell products to another consumer through a platform such as Amazon, Etsy, or eBay.

Consumers can even set up a C2C store on their personal website using online eCommerce tools like WooCommerce.

Consumer To Business

Consumer-to-business is the final type of eCommerce business model. This is the location where a consumer or individual entrepreneur sells their goods or services to businesses.

The Steps Of Creating Your Own eCommerce Website

Step 1: Choosing Your Business Model

The business models are already explained in a previous subsection so that they won't be explained here again.

Step 2: Clearly Understanding Your Business Goals

Establish your business objectives.

To begin, you must understand your company's objectives. First, you must identify your target market and customers. Identify your customers based on what they like and dislike, their shopping habits, their purchasing power, and so on. This information about your target customers will assist you in setting up features on the site and create a promotional campaign for your company.

How can I outperform my competitors?

To begin, identify your true competitors and the market in which they operate. Using relevant keywords, you can easily find this on search engines such as Google and Yahoo. Following the identification of your competitor, investigate what features and promotions they are using to attract and sell their products. Determine what technology and hosting platform they use to maintain and promote the site. Are they using a good eCommerce hosting platform or regularly optimizing their website?

This knowledge will assist you in configuring the features on your own website. You can determine which features will enable you to outperform your competitors and give them a run for their money. You can learn about some cutting-edge techniques that you can apply to your own website.

Create a project requirements list.

Make a list of the features and extensions you want to include on the site. Following that, create a detailed plan for the customization and extensions that you intend to include. Make a detailed list of the requirements and set a realistic deadline for completing these tasks.

Create an SEO strategy for your e-commerce website at the same time. Keep the following points in mind as you develop your website's SEO strategy:

a) Keyword research

b) Hierarchical navigation

c) URL enhancement

d) Appropriate categorization

Create a reasonable budget.

Once you've created a plan, make sure that your budget allows you to carry it out. If you have a $5000 budget, don't expect the developer to build an Amazon for you. Discuss your options with two or three good developers to see who can provide you with the most features and customization for the least amount of money.

Clearly define the design and technical aspects.

Check that the technical aspects are implemented thoroughly; otherwise, the entire project will lose its focus and momentum, turning it into a loss-making venture.

Examine the following technical aspects in depth:

a) Choosing the Best Ecommerce Platform

To truly succeed in the e-commerce business, one must select the right e-commerce platform. Fortunately, there are numerous e-commerce platforms available. You simply need to determine which one is best suited to your requirements. Suppose you are new to this market and have a limited budget. In that case, pricing may be an important consideration when selecting the right platform.

To begin, a start-up company should try some free platforms. However, if you have a large inventory, go for the reputable ones that can handle high traffic flow without difficulty. In addition, emphasize the template design. When a visitor first visits your online store, the design of your site has an impact on his mind. It will be difficult to attract more sales if it is poorly designed. So select a template that will give your store an elegant, professional appearance.

b) Create Product Pages That Work

In most e-commerce platforms, it is quite simple to add various products and product pages to any e-commerce store. It is critical to highlight your products by incorporating product pages into your e-commerce store.

c) Maintain the option of adding new features to the e-commerce site. Website

As we can see, an electronics e-commerce store is likely to have different requirements than a custom T-shirt store. As a result, most e-commerce platforms include various plug-ins that can be used to add special features to any website. On reputable platforms, you will have nearly limitless customization options. In contrast, cheaper platforms will only provide a few good plug-ins.

d) Maintain an active Search button on the e-commerce website.

According to statistics, nearly 30% of visitors use the search option on e-commerce sites to find the products they want to buy. As a result, it is critical to include "search" functionality to make the site more user-friendly.

e) Checkout and Payments Must Be Optimized

Before placing a purchase order, visitors to e-commerce sites must be able to create a user account. Once a user account is established, it can be used to follow up on future communications to promote newly arrived products. It also aids in the analysis of customer demographic data and the improvement of sales.

However, it is important to recognize that not every visitor will be eager to create an account. If someone wants to place a one-time order without creating a user account, that option should also be available.

Customers should be given various payment options for how they want to pay. However, the method must be simple and convenient for them. If the customer pays by debit or credit card, his personal information must be kept private.

f) Set up Analytics

Analytics has always been regarded as an important part of the success of business owners and commercial ecommerce platforms. If you want to compete for your customers' attention, or if you want to know how your ecommerce business's content marketing is doing, track your sales conversions, and better understand your audience, you must use effective and comprehensive analytics platforms. Google Analytics includes Webmaster Tools, which assist web developers in accurately and successfully calculating the return on investment.

These tools are widely used to monitor traffic, audience engagement, and conversion rates in both desktop and mobile versions of e-commerce stores. You can also examine which marketing campaigns performed well and which underperformed.

Step 3: Choose Your Business And Domain Name

Obviously, we cannot advise you on the best domain name for your specific website. We can, however, walk you through the process of making a careful, well-thought-out choice.

These aren't absolute rules – domain names aren't a science (you wouldn't have come up with "Google.com" if you followed these rules!) – but they are excellent best practices for most websites.

Select one or more keywords to represent your website.

There are numerous approaches to generating domain name ideas. You can simply base your site's domain on your own name or the name of your current business. On the other hand, a strong keyword is an essential component of a truly effective domain name.

Using keywords in your domain name is a good idea because you should use them in your content. SEO benefits can increase traffic by assisting your site in ranking higher in search engine results. Furthermore, keywords can assist potential visitors in better understanding your site's focus and niche.

Fortunately, researching keywords for your domain name is the same as finding effective keywords for other purposes. Tools like Google Keyword Planner and KWFinder can help you find terms with a high search volume but (ideally) low competition. Remember that you're looking for a keyword that represents your entire website and content in this case.

To explore your options, use a domain name generator.

Once you've narrowed down your keyword options, it's time to start thinking about domain names. While you can do this without any special tools, it is frequently difficult because you will keep running into names that have already been taken.

Using a domain name generator, such as Domain Wheel, will help simplify the process by brainstorming for you and ensuring that all of its suggestions are available for purchase.

All you have to do is go to the website and type a search term into the search box. As an example, imagine you're starting a dog training advice blog. You've decided on "dog training" as your keyword. Entering that term into the search bar yields the following results.

As you can see, there are numerous options available. If none of them appeals to you, you can try a slightly different keyword or scroll down for random ideas and keyword suggestions. We recommend that you take your time and come up with a few solid options so that you can narrow them down in the next step.

Based on a few simple criteria, choose the best domain name.

You are now ready to narrow down your shortlist and complete how to choose a domain name. A portion of this process is subjective – after all, it is critical to select a domain name that you like. However, there are some simple criteria you can use to help you make your decision.

For the best results, choose a domain name that is:

Short and to the point. Long and complicated names are more difficult to remember and type.

Ingenious and memorable. The more distinctive your domain name (while remaining clear and concise), the more memorable it will be.

Simple to read and pronounce. Avoid using numbers and special characters because they are difficult to understand when spoken aloud. It's also a good idea to ask your friends how to pronounce it, so you don't end up on this list by accident.

Finally, it is critical to carefully consider the Top-Level Domain of your domain name (TLD).

In layman's terms, this is the section at the end – .com,.org,.net, and so on. For many sites,.com is the best option because most visitors expect it and is the easiest to remember. However, finding quality.com domain names is becoming increasingly difficult, and users are becoming more accustomed to other extensions. So, if you absolutely adore a non-.com domain name, go ahead and purchase it.

On the other hand, TLDs that apply to a specific region should be avoided (like .co.UK or .ca). Google uses these extensions to help geotarget websites, so only use them if you truly want Google to geo target your site.

Get your new domain name registered.

All that remains is to purchase the domain name once you've determined the best combination of keyword-based name and TLD. The method you use will frequently be determined by the hosting provider you choose.

When you sign up for a hosting plan, you can usually purchase a domain name. You can also purchase a domain separately from a dedicated registrar or hosting company. If you've chosen a name using Domain Wheel, you can even go straight to a checkout page by clicking the View Details button.

When you're ready, use this Bluehost link to sign up for a hosting web account. It will result in two outcomes if you go through it: (1) It is an affiliate link, which means we get a small commission if you buy through it, and (2) it will get you a $2.95 vs. $3.95 discount. This discount will not be available if you visit the Bluehost website in any other way.

Of course, you are not required to obtain your hosting and domain name from the same provider. If you purchase a domain name somewhere else, you can still connect it to your new or existing site later. How this works will be determined by the web host you select. When making the transfer, you should consult your provider's documentation or contact their support team for assistance.

Step 4: Choose eCommerce Website Builder

There are numerous options for eCommerce website builders, but not all of them will provide exactly what you require.

When selecting a platform, consider your business goals, your product, and what is realistic for you to manage.

Consider the following questions when deciding on the best option.

Is it within your financial constraints?

When it comes to eCommerce site builders, pricing can vary greatly, so you'll need to consider how the platform's monthly (or annual) pricing structure fits into your budget.

Look for a platform with different tiers or packages so you can scale up your business and add more functionality as your needs change over time.

Is it possible to build the site within your time frame?

Suppose your site will require custom build-outs or developer work. In that case, you must consider the timeline for that work, as developer work can sometimes take several weeks (or even months) to complete properly.

Ensure that your back-end work and site build-out are completed well ahead of your launch date so that you are not scrambling at the last minute.

Do you like the available templates?

Examine the templates available on the eCommerce site builder you're considering to ensure that there are options there that your brand could use.

If you don't like the options on the site right now, you may need to have a custom theme created for your online store, which can be more expensive and time-consuming.

Is it compatible with the applications you require?

Integrations are what save you time and simplify the complexities of running your online business, so make sure the eCommerce site builder you're considering integrates with the software and tools you use.

That means you should double-check that your tools for logistics, customer data, email marketing, and so on will work seamlessly with the eCommerce platform you select.

Do they have the resources to assist you in building it?

As you build your store, you'll want to make sure you have solid support options that will make the process as simple as possible—nothing there's worse than needing help and not being able to get any answers.

Investigate the site builder's support options and see what general resources are available in the area of education and how-tos.

Can it do everything you need it to do without the help of a developer?

If you don't want to hire a developer to help you build out custom elements of your online store, make sure the eCommerce site builder you choose allows you to do everything you need it to on your own.

Make a list of the features you must have and ensure that the option you choose checks all of the boxes.

Spending time researching each store builder ensures that you choose the best one for your company.

Let's look at some of the best eCommerce website builders.

Step 5: Designing Your eCommerce Store

Your eCommerce website is your store, and it should be designed with your customers in mind. Your website should include information about your resellable products and services. Images, descriptions, prices, user comments, and ratings should all be included on your website to help prospective buyers. You must ensure that your website perfectly showcases your products so that buyers are never misled. You must make your web pages and website appealing because they are your brand's online representation to buyers.

Step 6: Set Up Your Payment Gateway

As an online business, you must provide customers with a variety of payment options. eStore builders, such as Shiprocket 360, include tools for automatically adding these features to your website. Payments should be accepted via credit card, debit card, online wallets, net banking, COD, and other methods.

Step 7: Secure Your Website By Installing SSL Certificate

All websites that transfer data over the internet must ensure that their connection is protected by a Secured Security Layer (SSL). The SSL certificate protects your website and fosters customer trust. Even Google now recommends that every website have an SSL certificate.

Step 8: Select Your Shipping Partner

Once you start selling products through your eCommerce website, you need to ship those products to your customers with the help of courier services. eCommerce logistics aggregator services like Shiprocket are the best option to go for. They give you multiple courier agency options to ship your product with the lowest available shipping charges so that you may have maximum profit value in your share.

Step 9: Test And Publish

Test Cases in General

General test cases for eCommerce websites/applications should be carefully and thoroughly written. Pay attention to the smallest details, such as the interaction quality of the homepage, the ease of navigation across product categories, or whether the product image enlarges upon clicks, among others.

- Some of the general tests that you should run on your website are as follows:

- Redirecting website links to the appropriate product/category pages

- Product, price, category name, price, company logo, and product description are clearly visible.

- Is there a relevant product listed specifically for each category on all of the category pages?

- If the total number of products listed on category pages is correct.

Test Cases on the Homepage

E-commerce website homepages go far beyond simple design elements. It is a powerful marketing tool commonly viewed as a clickable image or slideshow with auto-scroll redirecting your website visitors to specific pages.

The homepage serves as the profile space for your eCommerce website, so testing is essential at this stage. The following are the critical aspects of the homepage that the QA team must concentrate on:

- Is the page set to auto-scroll, and how frequently will the images be refreshed?

- Is it going to scroll to the next one when the customer hovers over it?

- Is it directing the customer to the correct page and product deal when clicked?

- Is the loading speed satisfactory?

Can the rest of the content, such as newsletters, banners, social media links in the site footer, and so on, be viewed easily?

Does the homepage look the same in different browsers and screen sizes?

Look for Test Cases

One of the most commonly used options in an eCommerce store is the search feature. Even the most extensive and intuitive category design can make it difficult for customers to find the product they seek. This makes it critical to test search features and make it easier for customers to find products quickly and easily.

The following are important things to test in the search feature:

- Is it possible to conduct a search based on the product name, brand name, or other criteria?
- Is it possible to sort results based on parameters such as price, brand, reviews/ratings, and others?
- What is the optimum number of results to show per page?
- Are there any options for navigating between multiple-page results?

It is important to note here that the customer can search for a product directly from the homepage or interior pages. As a result, your website search must be adjusted accordingly.

Products Suggestions

This is a critical component that is frequently overlooked in eCommerce testing. Following the client's purchase, there is a follow-up session in which customers are shown recommended items to purchase further. It is an important section to test because it serves as a foundation for long-term customer loyalty. Important things to look for here are –

- Check to see if the recommendations made to customers are for products that will be of interest to the client.
- Are the recommendations displayed on the page immediately after the client confirms their order?

Payments

Unsuccessful or failed transactions are among the most common reasons customers leave a website or online store without completing the purchase. As a result, payments are one of the most important test cases for running a successful e-commerce website. Here are some critical things to test on the payments page:

From start to finish, I'm looking into all of the different payment options.

- Can customers check out as guests to pay? Should they, on the other hand, register before checking out?
- Is it necessary for returning customers to sign in when they visit the page?
- Security checks for the storage of customer credit cards or other financial information. It is critical to take all precautions to ensure that user financial information is kept secure.
- What is the return page defined once the payment is made?
- When an order is completed, does the customer receive an order confirmation as an email/text message along with the order number?

Sixth. Shopping Cart

Shopping carts are one of the most important features of an eCommerce website, and they must be thoroughly tested. Customers can easily select and store multiple items in their shopping cart before purchasing them all at once. Some of the main test cases that should be included when testing a shopping cart are as follows:

- If the cart displays all items and their totals
- Add items to the shopping cart and continue shopping
- Taxes that apply depending on where you live
- Option to add additional items to the cart with an accurate total reflecting
- The ability to remove items from the cart
- Shipping costs are calculated accurately using various shipping options.
- Coupons can be applied if desired.

Product Details

Testing the product page is just as important in eCommerce website testing as shelving the goods available in a brick-and-mortar retail store. As a result, ensuring that these products display correctly on the site is critical.

Because the product page displays a lot of important information, such as the product description, image, specification, and pricing, all of this information must be displayed correctly whenever a customer logs in. Important things to look for here are –

- Product costing
- a photograph or photographs
- Technical Specifications (size, color, or variations options)
- Options for reviews and checking out
- Shipping specifics
- Options for delivery
- Details on in-stock and out-of-stock items
- Cases of Post-Order Testing

When we place an order on an eCommerce website, we have several options for following up on the purchase. As a result, testing the post-purchase functionality is an important aspect of eCommerce testing. Some of the most important post-purchase test cases are as follows:

- Checking to see if the customer can cancel or change the quantity of the order

- If the customer can review the most recent order and the history of purchased items
- Check to see if the customer can change their billing/shipping or other profile information.

Step 10: Make A Contingency Plan

Do you have a backup plan in place to keep operations running in the event of a site crash or a sudden surge in visitors? This is a critical issue because many start-up e-commerce businesses fail due to a lack of a contingency plan for unanticipated downtime issues. When customers are unable to access your e-commerce store, they immediately move on to the website of your nearest competitor. Furthermore, slow-loading pages harm your brand's perception. As a result, once your site has some traffic and sales, it must have specialized staff that handles all backup operations.

Chapter 4

The Basics Of Dropshipping

What Is Dropshipping?

Before you begin any business venture, it is always a good idea to know exactly what you are getting into. This way, you can make an informed choice with your time and money. Furthermore, knowing what you are getting into means that you can have an easier time actually creating your business because you know exactly what it takes to create success with that business model, which means you are more likely to be successful overall.

Instead of jumping straight into talking about how you can find products and suppliers and how to build business, we are going to the very basics. Even if you have already learned

about dropshipping in the past, I encourage you to read through the next couple of chapters with fresh eyes. Imagine that you have never heard of dropshipping before and that you are brand new to this concept, and that you are curious about how you can legitimately turn this into a business. This way, you can ensure that you have a strong understanding of exactly what it is that you are about to begin creating!

What Is Dropshipping?

To make it simple: dropshipping is a form of retail fulfillment where you have a company that is not responsible for keeping the products that it sells in its own possession. Rather than you being responsible for sourcing, purchasing, storing, shipping, and otherwise managing the products involved in your business, someone else takes care of all of this. When you sell merchandise from your store, the company responsible for your products will receive notification of the sale and fulfill the shipping for that product. They will also manage any returns or other product-related functions that need to happen for your business to run. You will only be responsible for paying that company for the products that you sell to your customers.

Inventory and the way that inventory is managed are the primary differences between dropshipping and standard retail companies. If you were running a standard retail company online, you would be required to purchase, store, and manage your entire inventory. This part of running a retail business can be incredibly time-consuming, as well as expensive, which is why many people do not begin retail businesses. If you can cut this part out, however, you can make the business far more sustainable while still having the capacity to earn a fairly high income from your business. This is where dropshipping comes in and is why dropshipping can be such a powerful way to create an income with your business.

Who Is Dropshipping for?

Dropshipping is an excellent business model for just about anyone to get started with. If you are brand new to online business, if you are an entrepreneur looking to spread out your income channels, if you are looking to get into entrepreneurship but have a low budget or testing out a niche, dropshipping is excellent for you to get started with. This business model does not require a large amount of capital to start with. It can offer a high payoff fairly quickly. Remain consistent in your efforts and follow a proven successful strategy from the start. You can use dropshipping to help you earn more money while also learning more about the market.

I want to elaborate briefly on the value that dropshipping has for helping you increase your market awareness and understanding. When it comes to getting involved in online marketing, it is important to understand the niche you will be selling and recognize how it works. Dropshipping can be a powerful tool because it could help you learn about a niche or a specific industry without investing too much capital into doing it. If you want to get started with online marketing in the health and wellness industry, for example, using dropshipping can assist you in identifying which niche will be the most accessible and profitable niche for you to get involved in. Then, once you have learned through dropshipping, you can either apply that knowledge to scale your dropshipping business or to launch alternative businesses in that industry.

When it comes to launching any form of business, it is essential to have adequate knowledge of that industry you are in and what it requires to generate success in that industry. If you take your time and use your dropshipping business properly, this can be a strong tool for helping you generate your own hands-on, proven analytics to teach you which part of your industry is most likely to be a success and how

you can make it even more successful. So, whether you are a beginner or someone looking to spread out into other profit streams, dropshipping is an excellent choice for you to grow an online business.

How Does Dropshipping Work?

You already know that dropshipping works by you not having to be hands-on with your company's products in any way, shape, or form. When it comes to dropshipping, you are only responsible for marketing products and offering a place for those products to be sold on. To help you have an even stronger understanding of what this looks like and how it works, however, I want to outline the dropshipping process in a step-by-step method for you so that you are clear on exactly what needs to happen for this business model.

The first step in launching a dropshipping business is choosing your

industry, building your brand, and creating a platform for your brand

to existing on. You can do this by identifying and selecting any major platform that works for selling products online or by creating your own with a web developer. Once you have created your platform, you need to select products for sale on that platform. Then, you need to create a sales funnel that will lead people to that platform to see your products and purchase them from you.

After everything has been properly set up for your business, you need to begin the process of actually marketing your business. For this part, you will begin by marketing to your niche audience according to your sales funnel. In most cases, the first step is marketing on social media so that you can build brand awareness and help drive attention and traffic to your platform. Your primary focus will be on marketing the elements of your sales funnels properly. In this way, you effectively reach the eyes of your target audience. f your sales funnel is designed properly, all you will need to do is market and create customer outreach. Then, your customers will find you and begin the process of moving through your sales funnel. If it is set up properly, you should not have to do much work beyond this as your customers will find you, land on your website, and then make a purchase.

Once the purchase has been made, a sale will be processed through your company platform. That sale will then alert your dropshipping supplier that a sale has been made, which triggers them to begin the process of packaging and to ship the product as per their company standards. They will also require a payment from you for each product that you sell, as this covers their costs of producing, storing, and shipping the products you are selling.

When the product reaches the hands of your customer, the sale is then complete. At this point, unless they request a return or are otherwise unhappy with their product, the process is done.

Why Should I Start Dropshipping?

As you are about to learn in Chapter 2, there are countless benefits to launching a dropshipping business that supplies many reasons for why you should get started. The biggest reason, nevertheless, would revolve around finances. Dropshipping is one of the least expensive and most passive income streams that you can start using online. If you are for someone who wants to increase your income but does not have a significant amount of money to invest in something to help you with this, dropshipping is excellent. Not only is it going to help you earn more, but it is also going to do so in a way that is perfectly manageable, no matter how much or how little time you have to invest in it.

Due to how technology has advanced over the years, you can use countless automation tools to make dropshipping even easier. Through this, you can successfully design and promote your dropshipping company in the smallest amount of time possible. If you want to scale your business, the increased time it will take to do so will still be minimal compared to other online marketing opportunities. For that reason, dropshipping is possibly the easiest and most lucrative business model for you to launch on the internet to date.

Setting Up A Mindset

Before you begin a business, it is always important to make sure that you are truly prepared to launch that business. Even with a businesslike dropshipping, which does not require as much effort as other businesses, is there still a certain mindset that you must obtain to ensure your success? Suppose you are not in the right mindset and don't have the right approach to your business, no matter how easy it is. In that case, you can almost certainly guarantee that you will fail in that business venture.

This chapter will explore what it takes to set up a successful business, how you can get the right mindset for obtaining success in your dropshipping business, and what you can physically and financially do to prepare for your business. This way, before you get started, you can guarantee that you are in the right frame of mind to succeed and to generate a significant profit through your business.

What Creates a Successful Business?

If you want to be successful in business, you need to be realistic, prepared, and willing. You need to be realistic about what you are going to face and what it is going to take so that you are well aware of what needs to happen for you to generate a positive result in your business. You need to be prepared always to have a plan for how you will proceed with things in your business, even if those things happen unexpectedly. Besides, you need to be willing to always work forward and find a way to guarantee a positive result for your business.

As an entrepreneur on any level, you are solely responsible for your business's success. No one will be as committed to your success as you are, so if you are not committed enough to make it work and see things through, the only thing that you can guarantee is your failure. You need to be ready to practically overcome anything you face in business and your life relating to your business. This means that if you are facing challenges regarding managing the finances of your business, you have to be prepared and willing to overcome that problem. Suppose you face challenges with your own motivation and consistency with showing up and marketing your business. In that case, you need to be prepared and willing to overcome that problem. No matter what you face in your business or life that impacts your business, you need to be prepared and willing to exceed those challenges to succeed in your business.

When asked, virtually every entrepreneur will tell you that their success comes from one thing: their mindset. Through having the right mindset, they have been able to identify solutions to problems that seemed impossible to defeat. With the right mindset, entrepreneurs have moved mountains to make their businesses successful, and they have had the energy to remain committed to seeing it through even when it seemed unlikely. As a result, they guaranteed their success by guaranteeing that they would not give up until they were successful. That is exactly what you need to be ready to do if you are going to start a successful dropshipping business, or any business, online or anywhere else in this world.

Leveraging Your Mindset for Success

Since your mindset is the most important aspect of your business, you should be prepared to leverage your mindset to achieve success. Many people who launch a dropshipping business are new to the industry. Therefore they are new to the requirements that it takes to make a successful run in the industry. It is not uncommon for drop-shippers to be individuals who are just starting out, who want to earn a side income, or who are looking to use dropshipping as a tool to teach them about the industry in general. In all of these scenarios, the mindset that is required to be successful may seem foreign and unknown. To help you get started, we will discuss the five most important mindsets that you need to have in place to help you generate achievement in your dropshipping business.

Be Fearless

Starting anything new in life can be terrifying, especially if you have high stakes involved in the new venture that you have chosen to begin in your life. Suppose you want to be successful in business. In that case, you will have to develop the mindset of being fearless to help you overcome any possible challenges or setbacks that you may face along the way. In business, there are countless hurdles that you are going to have to face, but the biggest one that people tend to underestimate is the fear of actually getting started. Many people put so much pressure on what it means to get into a business that they find themselves consistently backing out before they even begin, which will prevent your business from growing in obvious ways.

To develop a mindset of fearlessness, you do not actually have to *be* fearless. Instead, you simply need to behave like you are fearless. This means that you "feel the fear and do it anyway," as they say in the business. In other words, even if you are afraid and you think something is going to be challenging or possibly impossible, you take the time to actually go ahead and give it a try anyway. The worst that can happen is: you fail to make the results that you had hoped you would make, and so instead, you must use this failed attempt as a lesson to apply toward future efforts.

Develop Your Leadership Qualities

Even if you are the only "employee" that you have in your business, developing your leadership qualities is crucial to running a successful business in any industry. Teaching yourself how to become a better leader by improving your skills that directly affect your leadership means that you will have an easier time leading yourself and anyone else who may play in your business. Developing your fearlessness qualities is a great start, but there is plenty more that you can do to help you grow your leadership qualities to become an even better entrepreneur.

Leadership qualities are not only going to help you direct people in your business, but they will also help you motivate others and inspire them to join your venture, effectively helping you grow your business. Through leadership qualities, you will have an easier time acquiring customers, partners, stakeholders, and investors interested in supporting you in your venture. Without these qualities, you might struggle to really sell your vision and get yourself out of there, which can massively hold your business back.

Always Look for Solutions When it comes to business, you can guarantee that you will face several challenges. Look at any successful business that is well-known in its industry. You can guarantee that it has faced major challenges in reaching that level of success throughout its history. Even companies like Amazon and Coca-Cola have some noteworthy "failures" in their histories that they ultimately had no obvious solutions for. Even though the solutions they chose may seem like failures to their consumers, to the company, these solutions became major learning strategies that educated them on how they could become their best.

As an entrepreneur, the only way you will get through any challenge in your business is by being willing to look for solutions to the challenges you are facing. This means that if you had a launch that fell flat, you need to be ready to find a solution. If you have a supplier that is not pulling its weight, you need to be willing to look for a solution. Suppose you have customers that are consistently unhappy with a certain element of your business. In that case, you need to be willing to look for a solution. No matter what problem you are facing, you always need to be willing to look for a solution to provide your business and your customers with the best quality services possible.

Be a Grateful Person

Gratitude is a huge tool when it comes to growing your business. These days, many have earned the reputation of being entitled or ungrateful by failing to show their gratitude for their business and everyone involved in making their business a success. Virtually every industry is far more diverse than it once was, which means that the companies who behave this way rapidly lose business as people go elsewhere to companies that genuinely show gratitude for their business. If you want to achieve in your own business, you have to develop an attitude of gratitude and openly show it to the people who are partaking in your business.

When you run your business, always show gratitude for everyone. This means that even the suppliers who are helping you run your business should frequently be shown gratitude as it makes them more likely to offer you and your customers the best services possible. You should also show gratitude to your customers and everyone else in your business, and even your entire industry. The more you can show gratitude to people, both openly toward the public and privately behind closed doors, the better your reputation will be. People will see you as a pleasant business to work with, making it more likely for them actually to want to do business with you. As a result, you will be far more likely to grow and earn money with your business.

Surround Yourself with Inspiration

There will be many personal challenges that you face when it comes to handling a business, even if that business will not consume much of your time, like dropshipping. Every entrepreneur who has ever generated success in their businesses, regardless of whether or not the business was part-time or a full-time gig, have faced their own personal challenges that have made handling a business difficult for them. Suppose you want to be successful in running your own business. In that case, you need to be ready to overcome the personal challenges you face, making handling a business more difficult.

Surrounding yourself with inspiration is a great opportunity to help you exceed these challenges, as inspiration can remind you that there is a strong reason for why you are moving forward with your business. The more you surround yourself with inspiration, the more you will be lifted up and supported every time you need it. This means that you will have a much easier time generating success and scaling your business over time. Do not be afraid to join groups, follow inspirational people, and network with those who inspire you. The more you can surround yourself, the better you will do in the long run.

Financially Preparing for Your Dropshipping Business

On a practical level, before you launch your business, you will need to financially prepare for your dropshipping business. The average investor or financial advisor will tell you that you should be prepared to financially carry your business for 3-6 months with funds out of your own pocket so that you can manage the finances business until finances become profitable. Others will tell you that you need to be prepared to financially carry your business for 9-12 months, as this is a more likely timeframe for your business to generate enough profit to break even consistently.

On the issue of dropshipping, your startup costs and overheads are often quite low, so it is unlikely that you will need to save too much to launch your business with. However, having that extra money already put aside means that you are not going to have to worry about the finances of your business so that you can effectively keep your business up and running. At the same time, you work out the marketing aspect of things and drive traffic to your website.

With dropshipping, a saving of anywhere from $500 to $5,000 is ideal for helping keep your business afloat for at least 6 months while you develop your business. The exact amount you will need will depend on how you choose to get your business out there. One that

is run through suppliers, and a basic storefront will cost less, whereas one run through a company like Amazon FBA, where you have to purchase your own products and pay for storage and handling fees, will cost more.

The Pros And Cons Of Dropshipping

After hearing the definition of this phenomenal business model, you must be thinking how awesome this is. For entrepreneurs, it's much less demanding compared to conventional retail fulfillment models.

Let's take a look at a few of its main advantages.

Pros of Dropshipping

As a budding entrepreneur, you will find dropshipping a worthy resource that will not demand that much from you. You do not need to invest in a physical store, bear overhead costs, or stock products. Instead, you have an online marketplace where you buy products from third-party suppliers to sell to your customers.

Your role as a merchant is composed of gaining buyers and processing orders. In other words - you are a middleman or a mediator. Nevertheless, you can make tons of money if your approach is in the right direction to sell customer-centric products.

Dropshipping does not involve complicated planning. Here are some pros that can help motivate you to try this business model.

Low Order Fulfilment Expenses

In a conventional retail fulfillment scenario, you will need to have a potential customer's product available in your warehouse. Furthermore, the work will involve organizing, tracking, labeling, packing, and shipping the product to the customer. All of these can incur extra costs that you would have to pay.

Fortunately, you do not have to deal with all such operations in Drop-shipping. Instead of you, a third-party seller will take care of all this for you. Your only requirement is to track that your customer receives the order.

As a result, you save loads of money, giving you a commission in the shipment transaction that acts without your direct involvement.

Instant Sales

Conventionally, a merchant first needs to have all the products stocked in his/her warehouse to have them shipped to potential customers. After that, he uses a website for marketing the products. This is all done for you with drop-shipping, so there is no need to worry about storage and shipping.

With dropshipping, you negate this disadvantage by marketing a specific product instantly because you target products that are already available through another seller. That seller is the one who is managing all the products at his/her warehouse.

Low Inventory Costs

You barely require any storage for products, as you are selling products owned by someone else. That limits your expenses, allowing you to create a low-risk business.

Inventory costs are one of the highest expenses for retailers. You need to spend money for a warehouse and on the products. When you stock products, you often have to deal with an obsolete inventory, which forces you to reduce the stock. Otherwise, you might have to deal with limited inventory - causing a loss of revenue or stock that runs out. With dropshipping, you avoid all such hassles -

focusing only on increasing your clientele and improving your brand reputation.

More Diversification of Products

With dropshipping, you do not have to worry about marketing one or two products because you do not own any of them. You are offering products from various third-party vendors. This way, you have the benefit of offering a broad range of products to your customers.

More options for customers, higher opportunities to make more money!

And the best part is that you do not have to arrange any of these goods per se. Dropshipping lets you have the fruit grown on someone else's trees.

Risk-Free Testing

When you add new products to your inventory, you should note that a product is always at risk of failing to meet your target customers' standards. It often gets tricky to understand what customers expect. This can be a risky business if you are trying to market and sell products to buyers through a traditional eCommerce model.

In dropshipping, you do not need to own anything or make heavy investments in products. This allows you to test various products to see which one clicks with your customers. Once you find that sweet spot, you are good to invest in that particular product that your buyers like.

So, you save a lot without risking your valuable money.

Cons of Dropshipping

Of course, you will find drawbacks to this business model as well. Let's look at how dropshipping can be a disadvantage in the business world.

Complicated When Working with Multiple Sellers

Things are somewhat easier as long as you deal with a single seller who provides your customers with all their needed products, but when you work with multiple third-party suppliers, things get complicated.

With multiple sellers, you will need to keep track of every order, and each of these sellers may also have their own ways of processing, billing, and shipping orders. This can also lead to more time to ship products to your customers because you have to wait for the seller to ship the order to your customer. Uniformity helps your business.

Things can get even more complicated when the seller deals with multiple mediators like you, fulfilling orders for them simultaneously.

Limited Control Over Order Fulfilment

As sellers are handling all the shipping, and you barely have any involvement in the handling and shipping of the product to your customers. This often leads to delays in deliveries, mismatched orders, or low-quality products.

This can lead to a dissatisfied customer who may not be willing to buy from you in the future, as he deals with you directly rather than the seller. So, you have to bear the profit or loss of the final settlement.

So, one of the key areas to focus on when starting this business is to make sure you do business with the best suppliers in the market.

Less Profit Dropshipping requires you to sell more if you want to earn more profit. Naturally, you are adding a commission over the product price to sell it to a customer. The buyer could have got the same product from the same seller at a lower price if he/she omits your involvement from the deal. But this disadvantage can be reduced if you can sell items in bulk to your customers.

Poor Customer Service

You do not have to deal with the products directly, so you have no idea what the product might be like in reality. If there is a limited product description of the item, you will find it hard to answer customer queries.

Your only option in such a scenario is to forward questions to the seller, which can take additional time - leading to a frustrated customer. So, there is no real product experience. And the lack of information from you or on your website can harm your brand reputation.

Another common scenario is where you find out that the product you purchased for a customer is out of stock, even though it is shown as being "available." This frequently happens on platforms like eBay,

where sellers forget to give you an update that the product is already sold out. As a result, your customer leaves empty-handed with dissatisfaction and skepticism.

How to Deal With These Issues

Looking at the cons may make you feel skeptical about this model but think of it this way. If various large-scale companies use this model to process orders for their customers, it has to be a profitable model.

Dropshipping is more about making the right calls, choosing the right products and sellers, so your customer leaves and returns to get more goods from you with a smile. No doubt, you have to work less with the handling, processing, and shipping operations - but you cannot ignore the monitoring duties involved in this business model.

With the right approach, which you will find in the later chapters, you will have a better sense of judgment, improved business skills, and an idea of the platforms best suited for order fulfillment.

Choosing The Right Product

Dropshipping gives you a lot of freedom to decide which product to market and sell to your customers. But it's not as easy as it seems. You do need to invest your time and money in a product that customers can find worth buying.

To top it off, that product also needs to be of high quality. That involves you looking for top-grade suppliers who can provide you with the product at a favorable price. You have to consider several factors to ensure that your selected item is perfect for selling to your customers.

Once you have a clear idea of which products will be perfect for your dropshipping store, you'll start earning decent profits.

Let's dig deep into what to sell online.

Find Products That Sell

If you can find products that the customers will like, you can make a ton of money, but that process involves planning. Here are some points to keep in mind to find the right products that sell.

Choose What You Enjoy

The first thing that may come to your mind is to choose items that you like to get product ideas. You might want to choose products that are related to your hobbies, passions, or interests.

That's one of the easiest choices since you know what products you are searching for. You don't have to dig deep to understand what products are needed because you are already familiar with them.

So, you can do a couple of brainstorming sessions to find such products to sell to your customers.

Use Online Tools

Furthermore, you can use tools like Google Trends to search for the demand for such products in the market. Even if the product is not high in demand on the market, consider it a possible candidate because you will have less competition to market such a product.

Another way you can find products of interest to sell online is by browsing other online stores. It's a common strategy that many businesses apply to keep track of what their competitors are excelling at selling.

Using Social Media Shopping Platforms

You can find plenty of online stores on social media sites too. Just search for a potential product on Pinterest, Instagram, Facebook, etc., and you will have thousands of results lined up to give you various options.

Many merchants ignore these platforms because they don't know these sites' potential to showcase the trending products worth selling. Why not use all these sources of inspiration to figure out the products that your customers would love?

Satisfying their demand for the perfect product will help you generate more profit with your dropshipping business.

Look for Friend Suggestions

Social media websites are not the only places to brainstorm for finding the perfect products. You can also ask your friends the next time you go out to a social gathering. Ask about potential products that you have researched beforehand on Google Trends or another platform, and learn how they feel about them.

You can diversify your search even further by targeting individuals of all ages and groups. A broader range will give you a better scope of earning more.

As a result, you will have more ideas for your dropshipping business that you had not even thought of before.

Find Product Ideas in Everyday Things

There are so many products all around you that you need to take note of what people buy. Have a look around you and see all the amenities available to you. Think of which of the products present in shops near you could be a good fit for a dropshipping business.

You can choose products that people cannot live without. Or you can choose a product that helps simplify things. You can also choose a product that is not easily available to people nearby.

You have plenty of choices, and all you need to do is think clearly about the type of product you want to provide to your customers.

Travel to Help Brainstorm

People often find the best ideas when on the move. Maybe that could work for you too. Travel around and visit various places. Your mind will undeniably come up with loads of product ideas that you can sell to customers.

Make sure you are alert and thinking of opportunities that might knock on your mind's door at any time. Upon adopting such a mindset, you will have no trouble locating products that are a great fit to sell to earn a potential profit every day.

Making Decisions on Top-Selling Products

Once you know the type of products you can choose for your dropshipping business, you can start narrowing down the list. To do that, you will need to filter them based on the following factors. This lets you form a list of the best products to sell.

Keep in mind that your list is not limited to a shortlist of products. You can always add more to your list and see what is trending among customers, which may be profitable for you, eventually.

Importing products is fairly easy. The hard part is to know the items that will fill the buyers' needs and sell like hotcakes. The chosen products need to draw the attention of traffic to your online store. Once you find such products, you can market them and sell additional deals like upselling, cross-selling, etc.

Let's take a look at these factors.

Choose a Niche-Specific Product

The Dropshipping model has added more to the existing eCommerce platforms globally. The trend of niche stores has become popular prompting several entrepreneurs and organizations to use this successful strategy without competing with large-scale companies.

It does make sense - after all, you avoid fighting large online stores to grab the attention of potential buyers. When you target a specific niche, you omit product categories that are too general or broad. People are already exposed to tons of daily offers in general categories, which may obscure your marketplace from them. Instead, find niche items that big players skip or do not sell that much.

For example, you might find it hard to sell a normal cap to a customer, but you have more potential to earn profits when selling biking gear to a cyclist.

Focus on finding a niche that you would love to market. You will have a smaller audience, but every one of them has a higher potential to become a customer.

Narrow Down the Niche Products

Selling niche products is not a new concept in the eCommerce world. Many niche-specific categories already have a clientele that will not move to a new online store that easily.

So, you will need to penetrate even further to filter the niche products if you want to succeed in this business. When you narrow down your product list, excluding items that already have potential customers elsewhere, you increase your chances of targeting the right people.

A good way to filter niche products is by choosing somewhat new products. You will need to search among the various categories you are interested in and look for new inventions that emerged on the market but lack a potential audience. Marketing such items can create an authoritative platform for you to sell the items at a decent profit.

You can even target customized products in various niche categories, which are not easily available to the masses. The key is to minimize the product list as much as possible to have the highest chance of selling.

Look for Products That Customers Buy Repeatedly

As we mentioned above, dropshipping has a low-profit margin unless

you are selling products in bulk. Even with a low customer base, you can still earn good profits in the long run. The key is to choose products that customers buy repeatedly. This is a good strategy to sustain long term profits.

Choose products that individuals will keep buying over time and strengthen your brand's trustworthiness with them. Products like party supplies, clothing, and cosmetics are good examples of such products that people will need every once in a while.

You can target such goods to them with attractive marketing campaigns, triggering them to buy again. Use personalized emails and newsletters to let customers know that you have the products available and offer them a better deal, if possible. That way, you can create a long-lasting relationship with your buyers.

When a customer lasts with you for a lifetime, you barely need to sell in bulk because you have repeat profit coming your way every time you come up with something new.

Choose a Reasonable Price Range

The Online shopping world is like an ocean where you will find similar products listed by various sellers. These products will have various prices set for them. You will need to choose products that have an attractive price to lure customers to your online shopping portal.

Here, you can keep in mind a few fundamental rules: *Low price, high conversion rate.*

High price, more effort to find a potential buyer. You will need to choose a product that has a decent price range. The price should not be too high as it will not encourage buyers. Plus, such a product will barely give you any returns if you do manage to sell it.

Additionally, the price should not be so low that it feels like a cheap product or a rip-off. Customers are intelligent enough to tell what products may be of good quality based on the price.

The chosen price range differs demographically as well. Maybe a particular customer in Europe finds a $ 20 product expensive compared to living in North America. Do your homework while deciding the products specific to a targeted audience.

Choose Products That Deliver Faster

Customers often rely on services that can offer fast shipment of products. That's one factor that can help you decide the type of product you want to sell. Usually, smaller products can be picked, packed, and shipped faster than bigger ones.

Find niche products that your seller can quickly ship to your customers.

Choose Durable Products

While dropshipping, you will barely have to deal with the products physically. Everything, from choosing, packing, and shipping, is taken care of by the seller. You will want the product to reach your customer safely.

That's why you should choose products that are not fragile because you have little idea of how the product reaches the buyer. Sell products that can withstand heavy blows, the pressure of other goods, etc., while they are being shipped.

When you choose a fragile product, like ornaments, you have no control over the order fulfillment cycle. You do not want to live the memory of an unhappy customer frustrated over a broken product he received from your seller. Such a situation can lose you a potential customer forever.

Find something that is sturdy and durable for long transit journeys.

Finding Your Unique Supplier

Besides finding the right product to sell, you need to choose a good supplier too. Without a good third-party seller, you cannot effectively sell even the best product that you can find on the market for your customers.

A major part of your store's credibility will rely on the seller's operations, but finding a capable drop shipper is a daunting task. Often you will find it impossible to find one.

In this chapter, we will help you sort this issue out to find your unique supplier.

The first thing you should know is when to look for suppliers. Your search should be based on two possible scenarios:

You want both a supplier and a product.

You already have a product in mind, and you just want a supplier now.

Based on these situations, we will discuss the best way to find dropshipping suppliers.

Finding the Best Dropshipping Suppliers

Remember that a dropshipping seller will need to focus on the order fulfillment on your behalf. You must keep in mind the following points when choosing a potential seller for your business.

Research as Much as Possible

To find the best suppliers, you will need to do lots of research. If you have decided which products you want to sell to your potential clients, you should know which type of sellers to research online. Choose overseas or domestic suppliers based on the product niche you are targeting.

Once you list the potential suppliers, look deeper into their service insights such as delivery cycle, shipping charges, product charges, and customer service.

Talk to Them

You shouldn't just rely on research to know their fundamentals. You also have to speak to them to try to get to know their business acumen. Do they possess the integrity that you are looking for? Will they fulfill the orders for your customers with responsibility and commitment? You can often get an answer to such queries through a simple conversation with such suppliers.

Talking will also build a stronger relationship when you get to know each other. It is a way to build trust so that you and your supplying partner see things the same way.

Look for Sellers Who Can Send You Samples

Several suppliers do not mind sending samples to customers. You can use this facility to your advantage. If you have three or four potential suppliers in mind, you can ask for samples from each of them. Test the quality of the sample products, note the delivery time it took for you to receive the items, and other questions that you have in mind.

All the data will help you understand which supplier to choose for a healthy and long-term dropshipping business relationship. If you learn that the potential supplier is also supplying products to a competitor, get a sample from the competitor too.

Test both samples (one from the supplier and one from the competitor) to see if the quality is the same. You can also figure out if the competitor uses the same packaging that you will be using for your customers or some other custom labeling that makes their products and reputation stand out among their rivals.

Target Offline Suppliers

Most sellers who are present online may already be selling products to customers. That's why you should first try searching for potential suppliers offline. Maybe you have local stores that have little idea of how to market their products to customers online.

You can help such suppliers by partnering with them in your business. This gives you another major advantage: the seller will be near you, which will let you stay in contact with him/her more effectively.

Look for sellers who deal in niche products of interest to you and your business. You might have to provide them with a good presentation of your idea, so they will understand how they will profit from your business. Otherwise, they would have created an online portal of their own in the first place.

Pitch to Existing Online Suppliers

There are enough fish in the sea for everyone. Suppliers who are already dropshipping for customers can still collaborate with you if you are the right fit for each other. Choosing such a seller is beneficial because you do not have to teach such a person about the business. That saves you a lot of time - letting you move straight to the business if you strike a deal.

Online sellers who lack the skills and experience to manage a dropshipping business could be a good fit for your business. After all, you will become a potential resource to increase their business among your potential audience.

Find Manufacturers That Make Your Niche Products

Instead of targeting sellers, who might be buying products from other sellers, why not target the manufacturers? Companies that manufacture the products you want to sell can give you the most profit because you can remove all middlemen from the deal.

But such manufacturers need to have an effective order fulfillment operation, as you will just be placing the order on behalf of your customer. The rest of the process involving shipping should be available to the seller.

If you manage to find such a vendor-cum-manufacturer, you can get higher commissions on the products you sell for them.

Choose a Supplier with Professionalism

Professionalism is important in any type of business. Suppose you want to have a professional relationship with your vendor. In that case, you should both agree to terms and respect them throughout the cycle. Finding such a seller ensures that your dropshipping business is incredible hands.

A good way to judge a seller for credibility is by working with a seller who does not make you the first offer. You cannot entrust your new business into the hands of such a supplier who has to persuade you to be a part of the business and not the other way around.

Select a Punctual Supplier

Find a supplier that is punctual at processing your orders and shipping them to your customers without delays. Remember, it is not just the orders. Such a seller should also be readily available to answer any queries coming their way. Such vendors should stick to the scheduled shipment cycles to strengthen the relationship further.

Choose a Seller Who Is Nearby

This is not a necessity, but as mentioned earlier, it can be very effective. When you know that the seller is available nearby, you can always monitor their operations and make sure they fulfill your customer's orders with dedication and punctuality. Knowing your seller is a plus point that enables you to predict future relationships.

But that does not mean a seller located far away cannot give you a good service. It's a matter of how dedicated, honest, and trustworthy they are toward maintaining a good business relationship.

Avoid Choosing a Common Supplier

Suppose the supplier you choose is also supplying to 10 other eCommerce businesses in your area. In that case, it can be hard to emerge as a unique drop-shipping business. Try choosing a unique supplier who can ensure that you get a one-on-one order fulfillment deal so you can maintain your brand's reputation among end customers.

Find a Seller Who Sticks to a Contract

If a potential supplier asks to maintain a contract or an agreement to fulfill the orders, choose him/her. Such a seller will be beneficial for your business as he/she understands the importance of commitment. That means such a supplier will do his/her best to provide good-quality products for your customers - helping your credibility stay intact.

Choose Wisely

A potential supplier can help make or break your business. You have to make sure that such a seller either knows the business properly or is willing to cooperate to learn and grow his/her business with you.

A major part of the order fulfillment process is under the supplier's control. Finding the right one is important to make sure your business and reputation improve in the long run. Finding suppliers who are experienced in this field can be a great addition to your business model. Focus on searching for such sellers first.

If you cannot find them easily, look for the ones who have a sense of professionalism. That way, you know they are willing to learn and dedicate their products and services to your customers without any trouble.

The Right Platform For You

Once you have decided on the products to sell and the sellers to work with, you need to look for a reliable dropshipping platform. A dropshipping platform is a place where you sell services and products online through its sales portal.

Finding the right platform for your dropshipping business is significant for a successful business. To do this, you can choose two methods:

Build a website and self-host a dropshipping store.

Set up a sub-store on a renowned dropshipping marketplace or platform.

In this chapter, you will learn how to choose the right dropshipping platform, but before that, we will look into the different platforms you can use for dropshipping. We will also look at the factors you can evaluate when picking an ideal marketplace for your business.

Types of Dropshipping Platforms

Use Pre-Existing Marketplaces for Selling

Dropshippers often select such online platforms because they are resourceful. You do not have to make extra effort to run your business on such portals. All you need to do is set up your business account, upload your products, and start selling. Another plus point here is that you already have an audience visiting the platform to buy items - helping you market and sell your products effectively.

This method is quite easy, but you will still have challenges. For instance, such platforms provide limited customizations for your store. You will have to adjust to the features provided on the platform. You might be looking for features like site integration, fulfillment monitoring, or shipment tracking, but they may all be missing on such a platform. Some platforms have a limit on the number of products you can list on their portal. That, too, can frustrate you.

A restriction in personalization could be a significant downgrade to your branding. Moreover, working through a reputed platform hides your reputation to an extent. For instance, if you use Amazon to sell your products, your customers might tell other potential buyers that they bought the product from Amazon, as they will see Amazon's storefront instead of yours.

When you choose a pre-existing platform, you will also have to compete with other sellers on the same platform. Your pricing strategy will have to be top-notch so that you can attract customers to your storefront. Another downside is that you will have to pay a service fee to the chosen marketplace, which could be 10-15% of the order's price.

Setting Up Your Own Dropshipping Portal

If you open your own online store, you are responsible for everything, including complete control over your personal branding. Unlike pre-existing marketplaces, a customer buying from your website store will mention your website and the store name to other buyers. You don't need to set up your own website from scratch, but you might find building your own store exciting.

On top of that, you can use various plugins and tools of your choice to make your website's user interface interactive and appealing. There will be no limit to what you can add to such a portal which may have been limited to renowned portals. And the best part - you get to keep all the commission *yourself.*

Having your own marketplace with a personal website makes it easier to control the competition. Since your platform for dropshipping will be separate from other dropshippers, you can ask for better prices for your products.

The downside in using this platform is the time, money, and effort you will need to invest in building your website from scratch and coming up with a solid marketing campaign to build your traffic. You will have to work on

SEO and other marketing tactics to get a decent rank for your website on the search engine results.

Using a Self-Hosted Platform

If you are passionate about setting up a marketplace with complete personalization, have a decent amount of money to invest in your business, and have good technical skills, then a self-hosted platform could be a good option for you. This option is suitable for entrepreneurs who want to indulge in a serious money-making business for the long term. For such individuals, such a business is a full-time operation.

If you want to dropship full time and not just set it up as a side income, you can also get a domain name and hosting service for your business. With money to spare, you can hire web designers, content creators, and developers to help you to speed up the project. In no time, you will have your own marketplace ready at your disposal to sell niche products.

Self-hosting platforms will require you to host your website on a server that you own. Even though you have to spend a lot of money on building the site, your profits will give you ample returns to overcome the expenses.

As you gradually get to know the ins and outs of your business, you will upgrade it further with tools, plugins, and upgrades that will give you and your customers a better user experience. As a result, you can have a brand reputation that echoes your name in public without crediting other parties.

Even if you end up with issues, you can look for technical support at various online forums where other dropshippers like you meet and share their queries.

The Benefit: You will be the boss of your own business without any involvement from big eCommerce platforms. Your profits will be higher without a need to pay a share of your income to large conglomerates.

Using a Hosted Platform with Integrated CMS

If you do not want to involve website designers/developers or do not have the required technical expertise to set up a website for your business, choose a hosted platform with a content management system. Several sites offer plans where you can buy a domain and install eCommerce templates. You just have to fill in your personalized content, products, and descriptions to make your dropshipping business live for your audience.

This method lets you have the raw materials needed to build the site without using too much of your time. The service providers you choose for hosting will give you various pricing plans. You can select one of them to suit your budget. You can start with a basic plan initially if you want to save a little. Once your business shows lucrative results, shift to an advanced plan that can handle your traffic without ruining your site's efficiency. Talking about customization, most of the time, you will find what you want to add to your dropshipping website. A perk you get with this option is customer support from the service provider to help you whenever you are stuck with a technical issue on the site. This way, you don't have to rely on third-party forums online.

Furthermore, a content management system lets you add the content in readymade sections on the web page featuring various dummy products. Just swap existing images and text with a description related to your products.

The only disadvantage we can think of here is that personalization is limited. But you own the website in your brand name with a pre-existing template.

Factors to Consider When Choosing a Dropshipping CMS

Choosing an effective CMS can help your dropshipping business model work wonders. After all, the content management system acts as the backbone for your website to run with all the features you want.

Do pay heed to the following factors when choosing an effective drop-shipping CMS:

Provides Desk Performance - To know if your chosen CMS is a good fit for desk performance, check its reviews. You can also look for forum discussions regarding its efficiency. Also, you can talk to the CMS specialists as potential buyers to learn about the tool. An ideal CMS comes with a detailed guide that lets you learn about the system in depth. This information can include tutorials, blogs, tips, etc.

Features a Simple UI - A good CMS platform should not be complicated to use. It has to be user-friendly with a simple interface. You can go through the plugins and templates offered by a potential CMS platform for your dropshipping store. Browse and test through these resources as a user to understand if the CMS will be worth buying or not.

Offers Multi-Channel Support - You need a CMS that can support inventory data from multiple suppliers simultaneously. Your system should have a personalized layout to manage multiple vendors and the products they offer. For example, your content management system should offer unique product SKUs. Additionally, you should have an automated system that quickly adds products and their data for you.

Offer Miscellaneous Integrations - A content management system must have other integrated features like payment systems, email newsletters, and other services. Your platform should have multiple payment options for customers to give them flexibility in buying products from you. Compatibility can help increase your brand's reputation.

Features Responsive Layouts - You will need to select a CMS compatible with smart devices and desktops. You want the website to open responsively on mobiles, iPads, etc., as most people use their smartphones to shop for products.

Supports SEO - CMS systems integrated with SEO tools can be a big boon for your dropshipping store. SEO can help optimize your business site for an authoritative presence in your target niche.

Supports Blogging - It should also provide support for blogging if you want that to be a part of your website. When you add helpful content to your website in blogs, you motivate people to purchase items from your site.

Simplified Store Setup - Choose a platform for your drop-shipping business that offers help and support at every step of the process. Pay attention to the back-end operations your chosen platform can handle. After thorough research, comparing various platforms, choose the one that gives you the most features at a reasonable price.

Customer-Friendly - Select a website that has a customer-friendly setup. Such a layout should offer an appealing homepage, ample data for customers to comprehend about your products and services, shipping and refund policy details, and a well-categorized product catalog. Choose a drop-shipping platform that can offer you at least these features.

Additionally, your products should not be misleading. For instance, think of a situation where you see an appealing product advertised on a random website. You instantly feel like clicking and buying it, but you can't find a single product resembling the one you clicked on when you get to the website. That can be frustrating, right? Avoid following such a model and focus on genuinely improving the experience for your potential customer.

You can research the dropshipping websites of your competitors to understand what a good layout should be like. After that, see if the dropshipping platform you are about to choose offers similar customizations that you find attractive.

Supplier-Friendly - You need a dropshipping platform that can manage multiple vendors to ensure the consistent availability of products for your customers. The setup should be capable of adding new products to your website within a few clicks from various suppliers' sites. Overall, you want a streamlined process to manage all your products, SKUs, and other details without mixing up vendors.

Costs Reasonably - You can find content management systems available at various rates. Keep in mind your budget when choosing a CMS. Also, verify if the design templates and modules your chosen system supports are complementary or not. It all depends on how much you are willing to invest.

You will have to spend money on your dropshipping website, whether in the form of hosting service, CMS, or other tools that you want to integrate into the system. But, fortunately, these investments may not be higher than 15% of your complete dropshipping business budget.

Which One to Choose?

You should make a choice based on the objectives you want to accomplish in your business model. If your goal is to start your business as soon as possible, choose the pre-existing platforms with hosting services and integrated CMS. But, if you want to design everything based on your own creativity, buy a domain and design/develop the website yourself, or hire professionals to do it for you. To build your brand reputation, write blogs, optimize your website with the latest SEO strategies, and build a killer marketing campaign to promote your marketplace.

Choosing a self-hosted service can be exciting as you will be nurturing your website like your own child. But it can take time - invest only after properly thinking it over.

Factors to Consider When Choosing a Dropshipping Platform

Now, let's look into several factors to choose a dropshipping platform: Plugin Support - Choose a platform that gives you ample freedom to upgrade your dropshipping website. Maybe you want to add a new SEO-friendly feature to your website, or maybe you want to market your products to a new marketplace. Or perhaps you want to add a unique payment gateway like cryptocurrency. All such features should be readily available through plugins on the platform you choose.

Customer Tracker **-** Every order that your website fulfills should have a proper tracking process. Find a platform that gives you the feature to streamline the operations and manage your customer and vendor data with great efficiency.

Marketing Your Product

Even if you have the best website, CMS, and platform for your new dropshipping business, you will still need to get customers to reach your site. To do that, your top-priority tool is going to be marketing.

To sell more, you will need a solid marketing campaign with the potential to give decent returns in the long run.

Know Your Marketing Channels

There are plenty of marketing channels available nowadays. You need to decide which one suits you best for your requirements. The ones with the most potential can give you a decent return on your investment.

You cannot simply rely on the product ideas because if you do not have a potential audience to see your products, it will all be in vain. You will need to create a solid plan to sell your niche products because this is significant to becoming a successful drop shipper.

It becomes easier to sell stuff online when you have an appropriate marketing plan. It is marketing that will help you shine among your rivals. This section will help you familiarize yourself with the various marketing channels you can choose to create a lucrative business.

Note that some marketing platforms mentioned below will be better for selling specific products. You might not get universal results with just a single marketing channel. You will have to deduce which of them will be best suited for your dropshipping operations.

Facebook Marketing

eCommerce and Dropshipping work side by side - adding new ways of earning without the need of opening a brick-and-mortar store. It's a business model with ample opportunities to earn money online. One of the key features of the eCommerce business is that you can use your social media accounts, like Facebook, to help you make money.

Let's show you how you can do Facebook marketing for your dropshipping business.

People find it more convenient to find products they can buy at places that barely require any research work. Social media platforms like Facebook send tailor-made ads to viewers, showcasing what they might like to buy. It may sound similar to Google ads, but Facebook has the largest social network audience with billions. Starting an advertising campaign there might not be a bad idea.

Unlike Google ads, which viewers have to search for, Facebook ads are much more scalable. You share the ad with a couple of customers and, from there, it reaches millions if it is in demand. Facebook ads are planned with empathy to relate to them and find suitable products for their use.

The key is to master creating a Facebook ad. Let's show you how you can accomplish that.

Facebook Advertising in Your Dropshipping Business

If you are aware of online marketing, you already know the importance of SEO ranking. You may also be aware that ranking a website can take time. With Facebook Ads, you can head start your marketing campaign until your organic traffic starts building up - eventually increasing your sales. When you advertise your dropshipping business on a platform, your main objective is to generate enough traffic so that you can generate sales. You want to sell the products, for which your advertisement should be accurate enough to target potential buyers.

Facebook gives you the advantage of sharing promotional and sponsored posts in the same newsfeed. You can see the ads represented with the word "Sponsored."

How Does That Benefit You?

Your generated FB ad shows up in the newsfeed of a user's account on Facebook. Such an ad smartly blends in with the other normal posts, and, most of the time, the users barely notice that they are being shown a promoted post.

Psychologically, people ignore visual illustrations that are highlighted as ads, like Google ads. But, in the case of Facebook, these ads just feel like they're a normal post. If such an ad interests the user, he/she might click it and reach your website to browse for the product in the ad. Moreover, Facebook cleverly shows comments in such promoted posts, normalizing it even further to blend with the rest of the organic posts.

It's all about the first impression, which matters when selling a product successfully.

Facebook Ad Structure at a Glance

A successful Facebook Ad for a dropshipping business requires three stages:

1. Ad Campaign (for determining the objective of the ad)

Ad sets (for categorizing the ads based on your budget and target audience)

Ads (the actual ads that viewers will see) Let's discuss all three.

Campaign

This stage is all about setting up the primary goals for your ad. It's the starting point for determining how, where, and to whom your ads will be visible. Let's say you have two ad sets planned for your dropshipping site - one will market the complete website, and the other markets various deals on particular products.

With ad objectives in your campaign, you are generally accomplishing the following objectives:

Increase traffic to the website as much as possible Accomplish more likes on your posts and Facebook page Maximize your business audience Feature offers from time to time Boost brand awareness

Increase the sales of individual products listed on your drop-shipping website. When creating a campaign, you will have to categorize every ad carefully based on your objectives. You will also want to create an appropriate "campaign" for every category of products.

Ad Sets

In this phase, you are organizing the ads based on your target audience. You will categorize the audience based on their personal interests, purchasing frequency, recent events in their lives, their commenting and liking activities on various posts, their joined Facebook groups, occupation, and demography.

If you like, you can use "Facebook Audience Insights," a Facebook tool for helping you organize the process easily. It comes with the Facebook Ads Manager account.

For instance, if your niche products for sale are sports accessories, you will add details to your ad set for a 'personalized audience.' You will specify another ad set for individuals in the local area. You may also add another set targeting people who have made a purchase from your site in the past.

Ads

In this stage, you use organized ad sets and relevant data to your targeted audiences for marketing purposes. You create an actual ad in this phase that will be seen by the viewers from around the targeted demography. You will create several images, descriptions, and logos for every ad set you generate to see what works effectively for the masses.

Marketers often run ads using A/B tests, which means you create multiple ads with little changes made to each. You never know which ad may help you boost the conversion rate.

Targeting the Right People

It's a very simple logic: match ads with products that people find interesting or that are meant for them in a way.

You would not target your ads showcasing Mothercare products to a guy, would you?

Studies show that much dropshipping business fails considerably because the ads are not targeted toward the right masses. On the contrary, many businesses flourished because their ads reached the right viewers, who converted to customers eventually. So, you have to be careful when choosing the audience for your ads. If you have done ample research on your dropshipping niche, you will find it much easier to plan your potential audience.

Occupation, Interest, and Demography

It's quite obvious - a high school girl will not browse for the same items from your dropshipping website as a 70-year-old man. To make it easier, "Facebook Audience Insight" will be a helpful tool to allow you to organize your primarily targeted audience by their:

Occupation Age Interests (Facebook groups they have joined, pages they comment on, posts they like) Hobbies Purchasing Frequency Recent life events

If your niche products include cutlery, Facebook will autonomously filter out the users who recently purchased products related to the kitchen or anything related to that or who have had an interest in browsing them or are a part of a group in that niche.

It's like a laser targeting system that helps you look for potential customers without too much manual work. That's what you need for your dropshipping marketplace.

Geography

Focusing on geography in your ad campaign is important to narrow down the area of the target. The more concentrated your target circle is, the faster you will succeed at converting leads. The location should be one of the key preferences that will help you strengthen your Facebook Ad model for your dropshipping business. A potential customer located in a snowy region may find little use for an above ground swimming pool. You must target potential buyers within a specific geographical circle where you believe your products can reach them most effectively.

Custom Audiences

With Facebook marketing tools, you can create your personalized audience list to target your products. Creating a viewers' list will make it easier to find the right prospects with a high probability of buying your products. Such lists can feature past clients, individuals who contacted

you personally about a particular product, or people who reviewed your products positively. You can even target them with special offers or giveaways through Facebook Messenger.

Planning the Budget

Whether the amount is little or not, it matters when you spend it. In a dropshipping business, every penny is worth spending in the right way. To make sure that you do not waste your resources, we recommend planning everything slowly and smartly. Instead of spending it all at once, invest a small portion of it and see how your ad campaigns work.

Analyze your growth after the first investment and make financial decisions accordingly. Using the "Facebook Advertising Tool," you have the advantage of paying for the ads daily or in a lump sum, which is called the "Lifetime Budget." It depends on you how you want to proceed with the money you have for investing.

Make sure you are testing out new ways of improving your ads. You may lose money on a new technique, but you will not lose all of your money if you invest smartly. On the bright side, you will know that a failed ad has something wrong with it, which you can change and try again.

Learn the lessons from your creative ads and focus on the marketing part. Remember that this is the only main operation you have to do in a dropshipping business because the inventory management, picking, packaging, and shipping part will be the responsibility of your third-party supplier. Find new and effective ways to market your products so that you can get potential customers to boost your business.

Choose your ad campaign budget and utilize the money in the right direction with experience. Facebook will spend your invested money failing and has something wrong with it, which you can change and try again.

Learn the lessons from your creative ads and focus on the marketing part. Remember that this is the only main operation you have to do in a dropshipping business because the inventory management, picking, packaging, and shipping part will be the responsibility of your third-party supplier. Find new and effective ways to market your products so that you can get potential customers to boost your business.

Choose your ad campaign budget and utilize the money in the right direction with experience. Facebook will spend your invested money in two ways: based on the CPC (Cost Per Clicks) or CPM (Cost Per 1000 Impressions). With CPC, Facebook will generate ads on its own for you with the most relevant data to your product niche. With CPM, it will deduct money whenever your ad reaches 1000 impressions.

Facebook Ads Placement For Dropshipping

You are provided with five choices to place your ads on Facebook. These are:

- Desktop Feed Mobile Feed
- Instagram Feed Audience Network
- Right Column Feed (Desktop)

As per previous data, Mobile feed and Desktop feed are the most popular and have been known to produce traffic. You don't have to place ads in all of these, like the Right Column feed for desktops, which users ignore the most because it resembles an ad that online users are used to seeing on Google, etc.

Key Points to Include in a Facebook Ad

A successful Facebook ad should:

- Target potential customers who are willing to buy something based on the various categories.
- Build communication with your website viewers through email newsletters.
- Include a call-to-action to lead customers to your website.
- Not focus on the sales point of view directly.
- Offer giveaways and offers to your viewers.
- Utilize Facebook's carousel feature.
- Feature a price point.
- Share high-quality images and videos to build an audience.

Summarizing the Steps to Set Up Facebook Ads to Improve Dropshipping Sales

Investing in Facebook ads can do wonders for your business. To do that, you will need to set up a proper marketing campaign on this popular social media platform.

Here are the general steps you should follow:

Set Up Facebook Pixel

Facebook Pixel can give you a head start in your social media marketing campaign. It is an analyzing tool that lets you know how well your ads are performing on your website. Pixel lets you target your ads to the right people. You can go to the ads manager section of Facebook to "Create a Pixel" if you want to do it manually. Either way, Facebook can generate it automatically for you. Paste the generated pixel code on the web pages you want to analyze and let Facebook do its magic.

Create an Ad Account

To start making Facebook ads, you will first need to set up an ad account. With this account, you will make people aware of your brand and reach out to them even more. You will need to set up the Facebook Ads Manager to start with the marketing campaign. Furthermore, you will set up a target audience based on demographics and geographic area. You can optimize your ads here, noting the impressions, setting a target budget, etc.

Monitor Ads

You cannot just create ads and leave them be. You also have to analyze their performance to ensure a successful campaign. Take note of significant factors, such as Cost Per Action (CPA). Pay close attention to such features and make sure that their value is as high as possible so Facebook can optimize your campaigns gradually. Another factor to keep track of is the Click Through Rate (CTR). If this value goes down, it means your ads are not being watched. This number also needs to be high, so you will know your ads are getting views. A low CTR suggests that something is not right about the ad. You may need to make a few changes and analyze them again.

Optimize Ad Campaigns

As mentioned earlier, your work is not limited to creating ads and analyzing them. You will have to optimize them from time to time so that they show up as better numbers. It can be time-consuming at first, but you get better at this with time. Eventually, you will get perfect at deducing the right geographic areas and age groups to target with your ad campaigns.

Etsy Marketing

Dropshippers are always looking for new and efficient ways to market trending products and search for new products from various online stores, and Etsy is one of them. It's not your usual marketplace to drop-ship.

It's a place where you will find buyers investing in quirky products popular among the younger generation. With over 2.1 million vendors and over 40 million buyers listed on Etsy, you can consider this online platform to give you a lucrative market to promote your products.

Despite being a household name, Etsy's marketing for dropshipping is not for everyone. In this chapter, you will learn about Etsy, its pros and cons, and understand how you can market your products using this marketplace.

What is Etsy?

Etsy is a powerful eCommerce platform with a global reach. It usually engages in the trading of customized products. Etsy is one of the chief platforms to promote small businesses that want to sell their products.

This marketplace was established in 2005, and it has since generated over $200 million in revenue by the end of 2019. Despite the high number of buyers on this platform, many consumers and business owners underestimate this platform. But Etsy has potential that, if mastered, can result in really high returns in the long run.

Etsy navigation for dropshippers requires a little creativity with crafty entrepreneurship. You need to work hard to understand what is trending on the market to create a loyal client base that is ready to invest in your promoted products for the long term.

The Main Techniques of Etsy Dropshipping

Many dropshippers feel skeptical about using Etsy because it does not follow a beaten track. But that's the good thing about it. If you can avoid a majority of the competition, you can create a reliable eCommerce store for marketing your products.

Etsy uses two techniques for dropshipping:

The first one involves using another vendor's Etsy storefront as your own dropshipping store. But there is a limit to it. Etsy will not let you drop ship other Etsy seller's products if they are handcrafted by him/ her. Etsy considers dropshipping of handcrafted products illegal. So, you might not have a direct way of reselling such products.

Instead, you can create a storefront outside Etsy and sell the vintage and crafted goods from there.

The other method involves opening an Etsy storefront yourself and dropship from non-Etsy manufacturers. You can get the products from such non-Etsy stores and outsource them with your design to make them look like your handcrafted products.

These methods may seem complicated, but they are the best you can use if you plan to use Etsy marketing for your dropshipping business. This brings us to the section where we discuss the advantages and disadvantages of using dropshipping from Etsy.

Scaling Your Business Properly

This chapter will discuss an important parameter that will superficially work for taking your dropshipping business to the next level. Here, you will learn about the scaling process and ways to achieve it for business growth and expansion. It will only help your business when you are selling like crazy. You never know what is going to happen in the coming days.

Problems may hamper your business' sales, and that is where earlier planning of business scaling will work.

But!

Before we discuss ways to scale your dropshipping business, you need to understand the concept well. Some entrepreneurs even mistake `Scaling' with the word 'Growth,' which has a strikingly different meaning. However, we will discuss it in the latter part.

So!

What do you understand about the term "Scaling/Scalability?'

Scaling or Scalability is derived from the Latin word "Scale," which means "to climb by." In the business sense, scaling means taking a bigger leap toward revenue generation or growing your business.

In general, when a company scales, it soars high in revenue so that its losses make little impact. This way, the company does not only grow but will scale as well.

However, scaling a business is directly proportional to the capacity your business can endure. Therefore, before you actually plan to leap higher, you need a basic understanding of your resources, delivery capacity, infrastructure, and much more.

Once you have a thorough understanding of your capacity, you will need to create success milestones. Remember, if you fail to analyze your resources (delivery capacity, infrastructure, human resource, and other parameters), you may have a long list of failed deliveries, miscommunications, and worst of all, "unhappy customers."

But don't worry! We will teach you strategically about achieving scale without any loopholes. Before that, let me explain "how scaling is different from growth?"

How is Scaling Different From Growth?

We have often seen folks use the terms "scaling" and "growth" interchangeably. However, the two business terms have a distinct difference that every entrepreneur must understand. Learning the crucial difference will help in the strategic planning of business scaling.

Next time if some businessman uses "Growth" to represent the "Scalability" of his business in front of you, correct him right away.

Also, explain to him the definition that says when a company grows, it earns revenue by investing new funds in discrete sectors like promotions, human resource hiring, etc. In this case, the company's gains and losses are evened out without adding more values into gains. On the contrary, scalability defines the state of revenue generation where a company outpaces the liabilities and losses, adding more figures into its gains.

Having said that, while you plan your dropshipping startup and make strategies for its revenue generation, you must devise strategies to achieve scaling rather than growth. This way, you will not only increase the efficiency of your business but will and revenue safe when the market conditions are less favorable.

Steps for Scaling Your Dropshipping Business

It is worth considering scaling your dropshipping business. It will create a sustainable environment for long-term benefits unaffected by the ups and downs of the market.

However, you might be a little skeptical about how the scaling model will work. You may continuously ask the same questions, poverty and over, in your mind as to whether it will produce favorable results or not.

Well, I would suggest you stay neutral and optimistic about the scaling process. If you plan your branding, lower product acquisition costs, and take the help of online advertising techniques, you will be successful in creating the sustainable environment you yearned for.

Therefore, stop being skeptical about it and move with the process of scaling. You will require a few important steps to achieve this target.

Identify Your Business Milestones

When you identify your business milestones, you practically expect limited funds to invest (it is similar to every startup these days). Therefore, you need to invest well and hit the timeline before you run out of cash.

Here, it would not be wrong to say that you need to work out a roadmap to define milestones to hit within a given timeline. And even if your cash gets low, you can plan to fundraise and buy some extra time to achieve your milestones.

Further, it would be advisable not to think of valuation as a number that only increases. You can do a valuation by measuring your reduced risks and ease of doing business over time.

Focus More on Risk Reducers

Reduced risks mean healthy growth. If your business has fewer risks, you will be more focused on other attributes to increase growth rather than planning out to curb risks with passing the time. But do you have any idea about the risk reducers? Well, it is an attribute that can help your business reduce risks for investors and customers. You can take their help and plan out references for your potential investors and customers. The risk reducer will have data-backed proof for your potential investors so that they have faith in your brand for future investment, whereas, for customers, they will have data-backed proof of the benefits associated with your products backed up by previous sales.

Focus on Increasing the Sales

Your business solely flourishes on the daily sales your organization achieves. The more orders you get, the more sales are added to your lists. But! Where will you get the bookings from? Well, there is nothing very explicit to do. You can achieve a good number of orders by expanding your business reach, planning new marketing strategies, and promoting your brand.

Your ideal milestone should be equating the number of orders to the Average Rate of Return (ARR).

Identify Your Target Audience and Loyal Customers

Your business is nothing without your loyal customers. However, this process takes time. Therefore, you should first start gaining customers who will slowly become loyal to your valuable services.

Customer growth starts by channelizing your products to the target market. A few steps that will help midway are -

Selection of target customer base Start with product promotions Communicate your product's worth

All these three steps will help you fuel the product sales and gains. You can also deploy a feedback system to get considerable input from your users to study your buyers' behavior. Once you get the feedback, it will be easy to solve your sales issues and drawbacks.

Optimize Your Products According to the Ideal Buyers

Increasing sales by getting more consumers is an ideal process with time-to-time ignition. You can do this systematically by optimizing the pages where your product is promoted. If your customers have more inclination toward social media, run daily broadcasts and ask for social media reviews. You can also run blogs and publish articles about your products from time to time if you wish to have more of a presence among educated customers.

Keep the Number of Hired Salespeople in Check

It is good to have a robust human resource to back the company's process, but most entrepreneurs hire too quickly without thinking much about the aftereffect in the long run. Sometimes, hiring staff without any need can backfire and use up the company's founding.

Initially, you should hire staff that can help you with all processes. You can also call them "pathfinder and trailblazers." They will act as a salesperson or an administrator when required. Their multi-functionality will reduce the cost of staffing and add to more gains.

Have Good Control Over Supply Chains

Supply chains can make or break the backbone of any dropshipping business, which may adversely impact the scaling process.

As you are very well informed that the whole dropshipping business the model depends on procuring items on-demand from producers and delivering them to consumers; the slightest mismanagement in supply chains will kill the delivery speed and affect the appeal of your business negatively.

As you fail to deliver the items on time to your potential consumers, they will move away, which will affect your revenue and scaling process.

Therefore, before you work on bringing in more and more orders for your product lines, have firm control over your supply chains. Also, make sure the producers or suppliers you choose to work with have a wide range of items going with your brand, scalable inventories, and quality products.

Note: You will need to get in touch with all reputed suppliers to establish a successful dropshipping business in a particular industry. You will have to keep following the latest trends to see noticeable changes.

Also, changing your product suppliers will often limit the goodwill of your dropshipping business. You will have to remain consistent with the quality you have been delivering so far. It will drop the revenue margin very strikingly, affecting the overall scalability process if you cannot maintain your good reputation.

Achieve Customer Satisfaction

A satisfied customer is a key to a strong business. This rule applies to the dropshipping business model as well. If you focus on satisfying your buyers' product requirements, your dropshipping business will become trustworthy to your targeted niche. And suppose you even once dare to overlook the importance of listening to an unhappy customer. In that case, his negative experience could snowball the respect and goodwill you have earned so far.

Managing consumer grievances should be engraved as a core value not to affect the brand's value in the long run. The smallest spark online could damage the popularity and loyal customers you have earned so far, along with monetary revenues.

You can realize the impact by considering the example as explained here.

Suppose that you require a total of 2 weeks to procure a product and deliver it successfully to the consumer's doorstep. Now, let's assume that you made a total of 900 sales within the first two weeks and earned $50 per product sale. The earning per product has to be deducted for liabilities associated, like $15 for advertising and promotions and $10 for initial product purchase. After excluding all the liabilities, you are left with $25 as your profit (a happy time to celebrate, of course).

However, this happiness may not last long if your consumers did not like the product and want to return it. Your earned profit will roll over to zero, incurring you a heavy loss. You should note that you still have not calculated the shipping cost, payment transaction fees, bad relations with your supplier, etc.

Initially, this loss would be the most significant setback for the business scaling process. It could bring your enthusiasm and revenue down significantly.

Therefore, you should understand the value of customer satisfaction and take strategic measures to achieve it. The best you can do to pipeline 100% customer satisfaction is to run high-level and 24/7 customer support. The customers should be able to reach you at any point in time. This will avoid the issues of product returns and bad publicity.

Note: The customer support panel must be well-versed with customer handling etiquette so that unhappy customers are returned happily with a satisfactory experience.

Plan to Re-Engage Consumers With Retargeting Ads

Every seller, especially one from a dropshipping business, loves customers who come back for more. It is pure bliss that all your product lines are popular among customers, and they prefer to order from you.

But! How easy do you find it to achieve?

It depends on the seller. If you provide outstanding and quality products, your customer base will be more tempted to try every option available to them through you.

Again, the question arises - how will you re engage the old customer base to try new products? The answer is "with the help of retargeting ads and re-engagement." All you need is to get a strategic plan carved out for re-engagement and carried out under expert guidance.

You can take help from Facebook Ads, Google Ads, and social media broadcasting to produce more sales while maintaining the cost per campaign. Once you have successfully invested in re-engagement plans, you can, later on, invest in retargeting ads and increase the chances of brand visibility. Every time your customer base hears of your new product lines, they will be motivated enough to at least land on your website to learn about them. Rest assured, if your new product lines have that necessary charm to tempt them, they will immediately place further orders with you.

Develop Business Along With New Product Lines

It is good to have diverse product lines so that every customer returns satisfied from your doorstep. However, you cannot invest all your time and money in developing new product ideas. Besides, you will also need to develop your business and strengthen its base.

As a dropshipping business owner, you will have to work out all business elements and take them all together as a whole. Otherwise, the dream of a scalable business will go no further, requiring more investment and time.

Therefore, it would be highly beneficial to sit back and analyze your long-term business strategies and optimization opportunities. As you focus on business development, you will make new discoveries to expand and optimize your dropshipping business.

Note: Once you have a full-blown customer base and an ever-expanding sales graph, you may not have the time to improve or optimize any parameter associated with your business.

Achieve Low-Cost Customer Acquisition

Cost of customer acquisition or customer acquisition costs (CAC), as you may like to think of it, defines the investment made by you to attract customers and convert them into buyers. The low cost of customer acquisition will result in higher revenue, and fortunately, there is no shortage of such opportunities.

Suppose you go out onto the market and research low-cost ways to acquire customers. In that case, you will be surprised by the multitudes of options available like influencer marketing, affiliate marketing, content marketing, etc.

But the success will depend upon your products' reach and how efficiently it is marketed.

Note: CAC (customer acquisition costs) are variable and keep changing over time. There may be some methods that superficially worked earlier but did not even spark interest now. Therefore, testing will be required every time you plan to work out a new campaign.

Make Content Marketing Your Ally

Content marketing is a low-cost and authoritarian practice of customer acquisition and branding. The best part about this practice is that it works for every business sector and elevates brands.

Therefore, spare yourself some time for content marketing and make it your best ally for branding and customer acquisition. Fortunately, you will not have to create a different platform to make it work. It will utilize your website and social media handles to publish authoritative content and attract customers.

Utilize Web Analytics to Understand Your Audience

You cannot randomly sell your product to anybody. As a dropshipping business owner, you will have to collect data, analyze it and make data-driven plans to acquire customers and match their tastes.

However, the path is not that straightforward. You will need the help of web analytics to get familiar with your potential and ideal audiences to grow as a more popular and reliable business. With data help, you will understand your target audience and drive their attention toward your products using Keyword Research and Google trends.

There are many campaigns available like Facebook Analytics and Google Analytics to realize such goals. However, you will need a helping hand to predict the trends and help you sway customers.

Automate the Process of Order Fulfillment

Technology has rapidly changed the course of online selling over time. Receiving orders manually has lost weight, and automation has taken over. Precisely, this is the time when you either do it or lag behind with backlogs of orders and pending dreams of scaling your business.

Manual order fulfilling and scaling of business will not go hand in hand because manual processes are always on the riskier side, leading to errors and time consumption.

By automating the whole process of receiving the order and delivering it without being personally involved, you will save quality time spared for other tasks that remain unattended most of the time.

Moreover, it will be a significant step toward scaling your dropshipping business. It will reduce human errors, aid smooth order fulfillment, and help you achieve speedy delivery.

Automate the Process of Order Tracking

From receiving orders to dispatching them until the orders have successfully reached the customer, there are many instructions to be given and communication involved.

A person needs to alert any query placed regarding delivery and delivery timing, simultaneously looping delivery agents and managing the logistics. The whole process eats quality time which could be shifted to handling matters of much value.

For instance, you can make a portal where customers can log in their details and track their shipments online. This will reduce the need for your involvement, especially when dealing with transit-related queries.

Automatic emails, an online tracking platform or SMSs, will all create a huge impact on the automation process and aid scaling of your business.

Establish Good Relations With Supplier

A good relationship with your suppliers is a key aspect of a dropshipping business. You cannot succeed in the business without a fool-proof backing of suppliers offering low prices and bulk products.

Therefore, you should nurture a loyal and healthy relationship between all your big and small suppliers so that you can grow and expand with them.

A healthy relationship will pave the chance for better understanding and price negotiations when needed.

Note: Always keep in touch with your existing suppliers and maintain healthy communication. You can also prepare a monthly report of profits generated for them to keep you as a priority compared to other dropshipping sellers.

A Final Note On Orders

Orders with dropshipping happen differently from traditional eCommerce orders because the retailer selling the products has nothing to do with the products themselves. Despite this fact, it is still important that you understand the order process and its role in knowing how you can contribute to successful ordering experiences for your customers. This will also help you know what to do should any troubles arise to take all of the necessary actions to offset these risks and develop your business.

In this chapter, we will discuss everything about orders, ranging from dealing with them and what to do should any troubles arise during the ordering process.

Security

Creating a secure purchasing experience for your customers is crucial for them, as well as for you. In this day and age, people will rarely purchase online unless they feel confident that their purchases will be safe and that they are not investing their money into a scam. Improving security with your business by creating a secure checkout process for your customers is vital. The best way to improve order security is by using a strong platform like Amazon, Shopify, BigCommerce, WooCommerce, or any other well-known platform that can offer a strong checkout process. Refrain from using any unknown platform or one that seems to lack strong ratings, as this could result in your checkout process not being nearly as secure. Stick to well-known names so that your customers feel confident and safe when purchasing with you.

Fraud Issues

Using a well-known point of sale platform is a great way to eliminate fraud issues within your business. Ideally, you want to use a platform that will protect both yourself and your customers from possible fraud so that everyone feels safe as they use the checkout process. One big type of fraud that drop shippers face is credit card fraud. People will purchase products through them with a credit card and then order a chargeback through their credit card company to receive the funds back. This results in them having to pay for the product that the individual received and the individual receiving a refund for that product. In other words, they end up getting the product for free, and in many cases, they will sell the product so that they can earn a profit from it instead.

Protecting your company from credit card fraud starts with using a company like Amazon or Shopify, which has built-in features meant to protect merchants like you against this type of fraud. Certain shipping labels are used in these cases, which prove that the product was delivered, making it more challenging for the individual to request a chargeback.

Another thing you can do to look out for possible fraud cases is to check for common red flags that indicate that fraud could be taking place. For credit card fraud, the common signs include an individual who has different billing and shipping addresses, different names, strange emails that appear to be fake, rush shipping, especially on expensive purchases, package rerouting, or unusually large orders. These behaviors are common behaviors that occur when credit card fraud is happening and can massively impact your business. Suppose you notice any suspicious behaviors like this taking place. In that case, you can take action to prevent the sale from going through with your company.

One great way to minimize fraud and prevent chargebacks is by using a managed services solution, which is a company that pays attention to your incoming orders to prevent the risk of fraud. As your company grows, this may be an ideal opportunity for you to avoid being put at risk of these forms of fraud. A great company to consider should you want to use a service like this is ClearSale, which has built-in features that monitor your orders and flag possible fraudulent orders to prevent them from going through.

Product Returns

A strict return policy can prevent you from dealing with returns or refunds, but it can also drive people away from doing business with you since they cannot effectively test out your products to guarantee that they like them. Suppose your return and refund policy is too strict. In that case, customers may not want to purchase from you because they will worry that if they have a problem, they cannot receive any support with the problem they have faced. Instead, they will go elsewhere to another a business that has a better return or refund policy to purchase the same products.

Managing product returns and refunds with dropshipping can get messy, especially because the product has already been purchased from the wholesaler, making it more challenging for you to manage it. With dropshipping, you are dealing with the wholesaler's return policy, which means that if they have a strict return policy, you may have to refund the product and take it into your own possession, which begs the question: what then?

The best way to create strong return policies for your business is to choose suppliers who also have decent return policies so that you can return the product directly to the supplier. Alternatively, you can use a company like Amazon FBA who completely manages all returns and refunds for you and resells returned products to future customers. This way, rather than having returned products sitting in your own possession and with no way of really selling them, other than by selling them privately, they are all dealt with properly.

Shipping Issues

Running into shipping issues when you are a drop shipper can be challenging. Some suppliers take days or even weeks to work through these problems, making it even more challenging for you to deal with them. If you, as a merchant, take too long to deal with shipping issues, it can become a huge problem for your business. The best way to deal with shipping issues is to pick a carrier and then use that same carrier for every single shipment. For example, you could use the U.S. Postal Service or a shipping company like UPS or FedEx to do all of your shipping through. These services offer small businesses tools that they can use to calculate shipping on products, making it easier for those small businesses to offer consistent shipping rates. Alternatively, you could use Amazon FBA, which has its own shipping service that tends to be cheaper than going through your own private shipping deal with a company.

International Shipments

International shipments can be dealt with in the same way as domestic ones are; however, the pricing will be different. Offering a separate shipping option that provides prices for international customers is ideal to ensure that they are being charged for the entire shipping fee. This also ensures that you are not charging too much for the shipping fee, which could drive international customers away. Ideally, you should use the same carrier for international shipping that you use for domestic shipping, such that your shipping concerns are always easy for you to navigate.

Dealing with Out of Stock Orders

Most of the suppliers you work with will be supplying multiple drop-shipping companies just like yours. This means that accidents can happen where the products have to go out of stock and, on your website, it shows up as if they are still in stock. Unless you are using a company duo like Oberlo and Shopify, real-time updates are not often available, implying that you could run into issues around this.

Suppose a customer does order an out-of-stock product. In that case, the next best thing you can do is communicate with your supplier to determine when the product will be back in stock. If they are in stock quickly, you may be able just to let things go on as usual. Otherwise, it may be ideal to message that customer, offer any apologies for refund that part of their order. You may also consider offering them some form of an added bonus to make up for the inconvenience, such as free shipping on the rest of their order or a discount for their next purchase. In many cases, simply letting them know and offering a solution such as one of the ones previously mentioned will help you navigate these issues relatively seamlessly.

Inventory

While you run your dropshipping company, it is ideal for making sure that you have a system to help you identify how much stock is left in any given product to feel confident that you have plenty for your customers to order. If you use a simple company like Amazon FBA, this will be easy as Amazon updates your numbers on your dashboard to let you know how much stock you have left. Otherwise, you will have to regularly check in with your suppliers to see how inventory levels are going so that you know that there is plenty left for your customers to order. Suppose you find that inventory levels with your supplier are low. In that case, you can always set a product to "out of stock" even before it officially goes out of stock to avoid running into this problem.

Chapter 5

How To Make Youtube Channel To Earn Money Online

The Qualities Of A Successful YouTube Channel

With over a billion hours of videos being watched every single day, YouTube is one of the biggest powerhouses of content in the Wonderful World of the World Wide Web, and for a good reason. From random cat videos that keep your eyes glued to the screen from hours on end to serious, hard-hitting journalism about the ins and outs of politics across the globe, YouTube offers a wealth of video content for all kinds of viewers, no matter what you fancy. It's basically a one-stop solution for, well, practically anything, and it only makes sense for you to want to monetize this extremely powerful tool.

While it all may seem overwhelming, fret not - the good news is that you don't have to go through this journey alone. Itching to share your fabulous fashion sense with the rest of the world or clamoring for a piece of that cash-grab pie with your highly engaging live gaming feeds?

There's nothing wrong with dreaming big, but before you even get started on embarking on this epic quest, forget the nitty-gritty for a second and let's talk about what makes you, you!

YouTube sensations don't happen overnight, and not everyone is cut out to be the Next Big Thing. But here's the secret: successful YouTube channels actually have a handful of things in common, so let's get right to it:

Content That's Hard to Resist

With the myriad of things demanding viewers' attention on the Internet left and right, how can you make your own content stand out from the crowd? The most important thing you have to consider is to make sure that your content is really good. You have to be able to provide value to your viewers to grow your subscriber list. Your content not only has to be powerful and irresistible; it also has
to be creative, relatable, engaging, and authentic. If you're watching a video and you find yourself saying, "I wish I did that" or "Why didn't I think of that?" then that's a sign of good content. Try to incorporate it into your videos, but DON'T COPY.

Never underestimate your viewers' intelligence - they'll spot a fake from a mile away. This is exactly why you should make sure you actually know what you're saying and that you're not just regurgitating what other people are saying. Why should they listen to you? Why should they pick your channel over others'? You have to make sure that they finish every video with a sense of satisfaction and that your content is worthy of their time.

Religiously Uploaded Content (Perseverance is Key!)

You need to put in the commitment to upload videos regularly. Maintaining a YouTube channel takes a great deal of dedication; it's definitely not a cushy job where you can slack off. Creating content might look easy, but it's not! It needs your time and effort, especially if you plan to do this by yourself (and you really can, so don't sweat it!). You need to be able to think of great ideas, shoot them (as many times, again and again, to get your point across), edit, and do voice-overs. I'll be honest—be prepared to put in years of hard work without seeing any returns at first. If you're lucky, it can take a shorter time before you can rake in cash. YouTube success is not an overnight thing for most people.

Figure out how fast you can produce one video and come up with a timetable that works for you so you can upload content for your viewers regularly without putting too much stress on yourself. Doing so helps in creating a regular following, and since we all know that a human's attention span gets shorter and shorter as time goes by, uploading regularly reminds your viewers that you are still active and creating the content that they are looking for. Also, keep in mind that quality is always better than quantity—no use pushing out halfhearted content every day if it's no good.

Many YouTubers start off doing it part-time, and depending on the traction they get; they might end up doing it full-time. With all this, you need to remember to have fun and enjoy yourself. In the end, it's your passion for this that will keep you going.

An Ideal Niche Market

This move might seem like you're limiting your audience share, but this will actually make you stand out. You will be a newcomer in an already bustling and all-too-crowded world and will be competing against established brands who've been in the game for much longer than you have. It will definitely be a huge plus if you can find a niche to wedge your channel into. Find something unique - something you feel that only you can do. Add a twist or a flair that will make you shine. It might be your unique personality or a different way of showing things.

Love fashion? Look up and watch fashion YouTubers and research them and their style. Find something that they haven't done before and work on that. In the end, you will be marketing your YouTube channel yourself, and finding a niche market can help you cement a solid following.

Once you've already figured out where your channel stands, you will need to be able to stick to it. Ask yourself: will you be able to expound on this market share you've picked? Will you be able to produce content after content under this niche? Also, since you're just starting out, it will be helpful if you don't deviate because you need to build a proper following first. Only then can you think about expanding after that. Remember: people are looking for specific content that attracts them, and you want to make sure that the content that keeps them in place is yours.

Good Social Media Management

In this increasingly mobile world, a person's social media presence is like a ghostly third arm. It's invariably connected to you, and the same goes for your YouTube channel. This is why you need to promote your content consistently across all platforms and make sure all of them are linked back to your channel—it's a great way to put yourself out there as a YouTube creator. Maintaining a well-managed social media presence can really help you rake in those views, which may, in turn, boost your subscriber numbers. One of those coveted social media shares might just be the thing that could make one of your videos go viral—you just never know, right? Posting updates on your social media accounts can keep your channel fresh in the minds of your existing subscribers, while releasing quick, catchy snippets of your videos on Instagram, Facebook, and Twitter is a great way to attract viewers, new and old. You can also release teasers that could function as ads. Think of your channel as a brand. How will you market it? Do you start with a gimmick? Giveaways? How can you let your future viewers know that you exist and what they are looking for? How can you reel them in to watch your content and keep them there? Knowing what to post, when to post, and what to say has a huge effect. Remember: utilizing and managing your social media accounts well goes hand-in-hand with creating quality content for your channel, as these accounts will be an indispensable tool for promoting your channel.

Interaction with Your Viewers and Other YouTubers Alike

One of the awesome things about being a YouTube creator is having your own fanbase. As it is, your goal is to have a community of subscribers who would be loyal to your channel because they would be more inclined to watch, like, and share your videos. To do this, you need to work on getting more viewers to subscribe to your channel. Let's not forget that it will also help a lot if you have a relatable brand image or a friendly and engaging online persona. Remember: human interaction is important! You definitely wouldn't want people to feel like they're talking to a manufactured drone.

So, to get more subscribers, YouTubers tend to think of creative ways to reel the viewers in—and one of the key points is interaction.

Here's the deal: simply interacting with your viewers is one thing—we'll talk about that later. But interacting or collaborating with other YouTubers in the same field or not (if you can think of an angle) can also be a platform to boost your channel. Think about it this way: let's say YouTuber A has a million subscribers. Collaborating with YouTuber A means her million subscribers can also see your videos, and some of them, after watching your content, might subscribe to you too! Collaboration can mean a handful of different things, but in the end, it's basically a give-and-take situation. What would the other party get out of collaborating with you? It might be easier to approach a YouTuber with around the same number of subscribers as you do and plan a collaboration. You could come up with a 2-part video split between both your channels. This way, viewers would have to click on both your videos to get the whole story. You could also consider showing up on their channel for a one-off project. A simple shoutout can work wonders too!

Now, I'm not saying you can't do a collaboration with a more popular YouTuber. You just need to come up with an angle or offer them something you feel only you can give. Lay it all out on the table. Be confident and show them your passion for this. In the end, they are looking for that human connection—people tend to work with people they're comfortable with. Be yourself, but also be creative and try to think out of the box!

As for your interactions with subscribers, you'll have different ways on how to go about it. The main thing is to make them feel like they are part of your community or family. If you do this right, interacting with your viewers will make them feel that you're more approachable and relatable. It could be as simple as giving shoutouts or replying and liking to the first few comments that catch your eye. You could also give them options or ask for their opinions on how or what they want to see on your channel. And if you have a budget, giveaways can be good too! Make them feel that they are a part of your channel, as it should because it's their support that will keep your channel going. Lastly, always remember to treat them with love and respect because if you plan on doing this as your full-time job, both will go a long way

The Steps To Setting Up Your YouTube Channel

So, if you've figured that you already have the right qualities, then let's delve right in on how to start your own YouTube channel in detail.

What Your Channel Will Be All About

We already talked about this before, and you might already have a clear view of what you want to talk about in your channel, or maybe you have tons of topics inside your head, all competing against each other. This is perfectly understandable. Here are two things that can help you —research and experiments.

Now, this might sound boring but hear me out. This will help you narrow it down and save you from the future pain of trial and error—not to say that you won't be doing that in the future, but this will definitely lessen it and save you a great deal of time and effort. That aside, It will help a lot if you list out all the topics in your brain and do your research about every single one of them. To help you out with this, you can use Google Trends to figure out how your topics will fare (it's free!). It will show how many times your chosen topic was searched for on the internet, therefore showing how popular it is.

When in doubt, keep this in mind: all topics are interesting, however obscure they might be. Are you vlogging about how to make glass bubbles? Do you do tutorials on woodworking and other household handicrafts? Or are you just talking about random things like observing the microscopic ecosystem that grows over a week inside a jar with swamp water in it (this is actually a real thing—I kid you not)?

Whatever your topic might be, someone out there will still watch them. The important thing is to be consistent. If you're leaning a little bit toward the apprehensive side, you can always look for a topic on the upward trend or something stable. Keep in mind that trends will come and go. What's popular this year might not be popular in the next year, so try to think of something that will sustain the interest of your audiences for a long, long time.

You could also try various keyword search engines, but they will usually ask you to pay for their services. So, if you have the extra budget, you can use them; if not, you can always try the other approach—searching for your topic on YouTube itself. You will be hitting two birds with one stone with this one as you can see how popular your topic will be and how it would fare by looking at the number of views from existing videos. At the same time, doing this will also help you check out the competition—yes, my dears, competitor research.

Since you will be thinking of doing this to earn money in the long run or be famous (we can all be honest here), this is ultimately marketing yourself. You need to look for your "competitors." These are the people who are also doing videos under the same topic you're planning to do. Study them well and look at how they are presenting their ideas, styles, transitions, etc. List down the videos that are raking in those views. By doing this, you can have an idea of what works well. Here's another tip: you can write down all the ideas for a video you can come up with under your chosen topic. The more, the better! It's usually a bad sign if you're struggling to come up with a ton of ideas early on because you need to be able to churn out ideas to keep your channel going constantly. Remember what I said earlier: regular uploads.

Suppose, by some unfortunate twist of fate, you find it extremely difficult to develop a bunch of topics for your planned channel. In that case, it doesn't mean that you need to give up on it altogether. You can always try to look at things from a different angle. What exactly do I mean by that?

Say you're thinking about doing a topic on Photoshop tutorials. You roll your shoulders, flex your fingers, and—with a glint in your eye and an unstoppable go-getter attitude—you boot up your computer and check out all the videos on the topic on YouTube. To your complete and utter dismay, you find that there are already heaps and heaps of tutorials on Photoshop existing out there—how can you possibly make yourself stand out?

Before you rage-quit and flip tables and call it a day, think about a different angle. Is it possible to narrow down such a broad topic? What kind of Photoshop tutorials can you do? Is there anything new that you can offer viewers that the others haven't offered them before? What else can you bring to the table?

Here's an example: you can try to do tutorials that focus solely on removing the background from an image. You can narrow this down further by teaching people how to use a plain white background instead. Focus on this more by targeting small business owners who want to open their own online shop—how can they make the images of the products they're selling as clean and clear as can be? How can they retouch product photos in the best way possible to boost their sales? This alone can already help you grab a huge chunk of the YouTube market, so to speak. You can even inject a little bit of humor in there if you want to—funny stuff is always a hit!

You're only starting out, after all, so let's be honest: you won't have a lot of subscribers right away from the get-go unless you're already some kind of celebrity. But if you already have a channel lying around, you can try posting videos of your chosen topics to get its feel. The more you post, the more you'll learn. It may be a huge cliche, but experience is always the best teacher!

Create Your YouTube channel

Now that you have figured out your topic, congratulations! Next comes creating an account. Since YouTube is owned by Google, you will need to have a Google account on hand. Using that Google Account, you can now sign in on YouTube and create your channel. You have two choices: create a personal channel where only you can manage your account or a channel with a business or another name. Most YouTubers looking to earn money would pick the latter.

For the personal channel, it's fairly straightforward. Log in to your YouTube account—desktop or mobile; it doesn't matter—then try anything you would do with a channel, like uploading videos, commenting, or creating a playlist. If you don't have a channel, a prompt will ask you to create one. Enter your details and confirm. Viola! You have a channel! Most people will have this personal channel already if you regularly use YouTube. If you have a playlist and can post comments, then you should already have a personal one. Next, for the business channel—again, most YouTubers will use this—this can have multiple people managing it, which can be useful in the future when your channel has grown big enough to necessitate hiring a team to work for you (still, some people do this so they can just concentrate on creating content). For this, you can use your Google Brand Account, or you can just create a new channel. Using your desktop or laptop (it's easier), go to your channel list, which can be found under "Settings" on your YouTube account. Click "Add or manage account," then select "Create a new channel." This will direct you to create a Brand Account. Fill out the Brand Account name, click "Create," and you now have your own channel. Yay!

Customizing Your Channel

When you click "Your Channel," which is located on the drop-down drawer when you click your profile pic on the upper right-hand side of YouTube, you'll be brought to a page with two blue buttons on the upper right-hand side of the screen that says, "Customize Channel" and "YouTube Studio." In "Customize Channel," under "About," you can add your channel description. You can add keywords here to improve your chances of being discovered by viewers when searching for videos. You can also add links to your other social media platforms or your website (limit of 5 links). This will appear overlaid on your channel art. By clicking "Add channel art" (this is the blue button in the middle of the banner up top), you can change your channel art. If the blue button is not there, click the pen icon on the upper right corner, then click "Edit channel art." The channel art appears as a banner on top of your page when people visit your channel. Change your profile pic here, too, by clicking the pen icon on the upper left-hand side of the banner.

If you are using a Brand Account, when you click "Change," you'll be notified that your profile pic comes from your Google Account (since it is connected). If you click "Edit," you'll be brought to your "About Me" page in Google. Clicking the pen icon on the right of your default profile pic will allow you to change it across your Google Account. We'll talk more about these later when we're talking about branding your channel. You'll also see a gear icon on the upper right-hand side of your screen. Clicking this gear will bring up your channel settings. Under "Customize the layout of your channel," click "Advanced settings," and you'll be able to set your country and enter keywords about your channel. Type in everything you can think of that could describe your channel. What is it about? Will it be about you? Vlog? Reporting? Trivia? Pets? Food? Hobbies? Under what topic? Like you did in the channel description, these keywords will help people discover your channel. The more detailed it is, the better your chances of showing up on YouTube searches will be, which can mean more potential viewers! Think of this like the hashtags you put on Instagram but without the hash key. Also, on this page, you can link your AdWords account and your Google Analytics property tracking ID. After making the necessary changes that you deem fit, click "Save," and from here, you can access YouTube Studio by clicking "Return to YouTube Studio." It is inside this YouTube Studio where you can handle all your creator stuff. This will be your main control room as a creator, as you can do almost everything from here. Your videos are here, your comments are here, and your analytics are also here. You can also put subtitles for your videos on this page (international audience, yay!).

Most importantly, you can also go to monetization from here. Everything and anything can be accessed on the toolbar on your left, which is under your profile pic. Another gear icon at the bottom of this toolbar allows you to access "Settings." You can further customize your channel here. Explore! Tweak! Not every channel is similar.

Up next is verifying your channel. Verifying your channel is actually easy—you just need to click your channel icon, click "Settings," then click "Channel status and features." Click the blue button beside your icon that says "Verify." I suggest that you have your mobile phone within reach during this stage. After you have entered your country, you will receive a call or a text from YouTube, whichever you picked from the options.

Enter the code you receive and click "Submit." When you go back, you'll see that your account is now verified. Now that you have your channel verified, you can now post videos that are longer than 15 mins, live stream, and appeal content ID claims. You can also set custom thumbnails for your videos! This is also as important as your title because humans are visual animals and images register faster in the brain before text does. So, making a custom thumbnail will be a great addition as it can help you attract viewers. Also, it will be good for your branding.

The verification badge, on the other hand, is an entirely different thing. This is when you see a checkmark next to the channel's name. You can submit a request for this when you have finally reached 100k subscribers. And finally, after all that clicking and tweaking, you have just finished setting up your own YouTube channel. Hooray!

The Process Of Branding Your Channel

Suppose you take the time to observe the more popular YouTubers. In that case, you'll notice one thing: everything has this same look going on, from the logo to the video thumbnails. This makes everything look cohesive, clean, and easy to understand. You'll find that you can identify the YouTuber by just looking at their video thumbnail. This is the recall you are going for. I'm going to walk you through just that so you'll have a style or branding that people will associate your channel with, which, in turn, will make you more memorable.

Think about it: when somebody mentions McDonald's, what image comes to mind immediately? You may think about the golden arches forming the letter "M" or the vibrant red and yellow colors that pop out from afar. You may think of the delectable smell of crispy fries that wafts in from the kitchen as you step through the doors into the fast-food joint. You may instantly recall fond childhood memories of sweet sundaes with the family or milkshakes with your best friends. You may think about biting into the juiciest burgers in the company of loved ones or of the din and the laughter as you're surrounded by lively chatter and pleasant conversation.

Not all of these feelings that the mention of the brand evokes are intentional—all these may just be from your own subconscious even if you're not aware of it. That's the ultimate power of branding—it helps you build an image in your mind and your heart about a certain product or service. Wouldn't you want your viewers to experience the same kind of epiphany whenever your channel is brought up in casual conversation?

Let us start with the basics. To build the brand identity of your channel, start with your voice. This is the image you would want to project with how you present your videos. Since you already have your topic in mind, think of how you would want to present it. Are you going to be friendly? Relatable? Informative? A serious journalist? Quirky? A moody hipster? This should definitely be in line with your own personality—remember that people can spot fakes from a mile away, so it pays to be genuine.

Next, pinpoint your target audience. What do they like? What's their age range? What are their online habits and behaviors? At the risk of looking like a creepy stalker, you have to find out their likes, dislikes, quirks, and idiosyncrasies. You should be completely obsessed with your audience—you can find all of this info from conducting market research.

Market research is a different beast altogether—even huge multinational conglomerates hire third-party marketing teams to study their markets and figure out what makes them tick. So that you can have a better picture of just how important market research is to any brand, think about the big video game company Nintendo. When video game consoles were just starting out,

Nintendo was unstoppable—the corporate giant was dominating the industry with more than 90% of the market share at the time. Sega, an up-and-comer in the same industry, had all-too-many difficulties trying to keep up. While Nintendo was only operating on 8-bit with their games, Sega already had a much more technologically advanced 16-bit engine working for them—but an even stronger physical performance in the console could help them outmatch Nintendo in any way.

To get them out of this dilemma, they came up with a strong marketing plan on how to make their mark in the industry—and this involved a solid foundation of market research. Since Nintendo was marketing their games to a younger market, Sega decided to grab the untouched market share from the older bracket: the teens. To do this, they studied the behaviors, wants, and needs of teenagers. How do they behave? What's important to them? How do they want their games to play out, and what factors will keep them hooked onto their screens?

The answer came in the form of Sonic the Hedgehog—a fast, reckless, bright blue hedgehog with snazzy sneakers and a no-nonsense attitude. Sonic's personality—and his penchant for speed, speed, speed—instantly clicked with teenagers way more than Nintendo's cute plumber Mario did during that time, and this helped Sega rise to the top and be at par with Nintendo. Of course, Nintendo is still going strong today, with Sega lagging behind again, but I digress—that's a different story altogether.

The point is that knowing your target market inside and out is crucial to any branding strategy, no matter which industry you may be in.

With these in mind, it's time to start building your brand identity. You will be using your voice to reach out to your target market, playing into what they need and want. Start with your logo or icon; you can also use your photo if you're a vlogger. Extend this branding to your channel art, video thumbnails, and video watermark if you want one. To make things easier, you can try making a simple style guide, but a mood board will also do for starters. Pick a color palette or set of colors that you will use it in your channel, and if you are a business, you should use your brand colors. Your fonts and graphics should also work well with your branding.

Build a Template for Your Video Thumbnails

Some people, like the vloggers or the more modern channels, go for a more organic approach. Some great examples of this style can be seen on Vox and a few vloggers like "best dressed." Vox, of course, has an art director to do those thumbnails for them (you can do it yourself, too, if you're good with design and layout). The result is an effectively recognizable video thumbnail anywhere. This is effective because when their videos come up in your recommended feed, you know that it's one of theirs, with those bright yellow accents and boxed headlines. They have since branched out from this style as their channel grows with different subtopics, but they still have this same set of colors and graphic style that proudly screams their brand. On the other hand, some vloggers have a more scrapbook-y feel, with hand-lettered fonts and quirky cutouts to boot.

Make Use of Palettes

To do this, you'll need to have a few fonts and graphics you'll have on hand that you will use to play around with. The trick here is to have a specific color palette and to adjust the color of your pictures. This is called color grading.

Take a look at some of the more famous Instagrammers. If you look at their feed, every shot has more or less the same feel: cool or warm, high saturation, or black and white. Some even go as far as choosing a predominant hue. It makes everything pleasant to look at, cohesive, and not scattered all over the place. This can be done easily with Adobe Photoshop or Lightroom. You can also use simple photo editing apps on your phone like VSCO.

Lightroom and VSCO have presets that you can use as a beginner. This same color grading trick can also be applied to the video thumbnail and your video itself. When done right and you have your own style, it can greatly help with your brand recall. Just pair it with your selected fonts and graphics, and you'll be good to go!

Consider Uniform Templates

On the other hand, some people would prefer to make their thumbnails look uniform all throughout. This is easier to pull off as you will have a set template that you can use over and over again—you'll only need to change the text and pictures every time you have a new video. This way, you won't be pressured to think of a new layout every time you have a new one coming up (again, don't forget to color grade your pictures). This works best if you are leaning more on the informative or minimalist side.

Let's say you do video editing tutorials on your channel. This style should be distinctive and informative enough that they will click on your video if they search for a topic and your video comes out, especially if you already have the brand recall you're aiming for. Another thing to note is that your target audience, once they are in your channel, will definitely have an easier time browsing through your videos to look for the specific topic they are interested in.

Keep Thumbnails Consistent

With either approach, especially with the organic style, try to keep them consistent and looking like they belong with each other for easier recall. You will be building your brand recognition, after all—plus, you will also be making it easier for yourself when you are branching out to other subtopics in the future.

Let's say you are originally doing DIY videos, but somewhere along the way, your viewers decide they want to see more vlogs from you. With consistent brand guidelines set, you can easily incorporate your original thumbnail layout and make a new one using the same elements by shifting the color or the elements around a bit.

Another way is to make a mini "flag" or "call out" to be placed on the edge of your thumbnails. Want to position it at the bottom? On the right? Left? It's totally your choice. This is to separate them from your original content, and don't forget to utilize the playlists too! Subdivide the categories of your videos to make them easier for your viewers to browse through. You don't want your subscribers to slug through the hundreds of your videos just to find that series of DIY cat mufflers you did a year ago. If it takes them forever to find it, they will quit mid-scroll.

I know this is a lot to go through just for video thumbnails, but I just can't stress this enough. The thumbnails will be the first thing that attracts the viewers' eyes when they're searching for videos. Well-made and eye-catching thumbnails are sure to attract viewers. Think of this as the packaging of goods in grocery stores—companies and brands invest large sums of money to have designers create good packaging. It might seem unimportant as you throw it away after opening or using it, but ultimately, it's what attracts your eyes first. It makes you want to touch the product, inspect it, and hopefully even buy it.

It's the same thing with your thumbnails—viewers scroll really quickly sometimes, and you only have a few seconds tops to attract their attention. The solution then is to make it interesting enough—or, for lack of a better term, clickbait-y enough—to make your viewers click and enjoy.

Don't Forget Your Channel Art

To recap, the channel art is the banner up top that viewers see when they visit your channel. A lot of first-time YouTubers make the mistake of making this image static and never changing. Don't make the same mistake! Treat this as a billboard for your channel or your calling card. As I mentioned above, you can put links to your website or your various social media accounts here. If you use

YouTube as a social media extension of your business or your brand, this space will be a great place to advertise it.

You can even advertise how often you post new videos here. Use catchy phrases like "check back every Tuesday at 8 am for new barbeque videos!" or something to that effect. You can also use this space to promote your new videos, a new channel, or a new segment. I suggest changing this banner every time you have something new happening in your channel. Are you a business with a new product? Advertise it here. It's basically like a billboard that you don't have to pay for, so utilize it well.

Brand Your Content, and Yourself

Now let's talk about the branding for your videos themselves. The things I've been talking about above are meant to catch your viewer's attention. We can consider them as the outer layers of your content, like the gift wrap or packaging. This time, we will talk about the meaty, juicy inside, which means your videos or the content itself.

You might be thinking, "Does branding apply to this stuff too?" Why, yes, it does! I mean, you'd definitely want to hook passing or casual viewers with your amazing content and turn them into active subscribers, right?

We live in a consumer-driven world, and everything is out there to please or attract you and make you want or need things. Corporations invest millions and strive to turn your "want" into a "need" and to make you buy their products repeatedly. Case in point: Nike comes up with new basketball shoes that they say to make you jump higher. Sneaker-heads and basketball enthusiasts "need" it, which translates to a sale. See the trick?

Now, I'm not saying you should go ahead and trick your viewers. Remember when I said that they spot fakes from a mile away? Plus, they will talk about you in forums, and word spreads fast on the World Wide Web (alas, such are the woes of becoming famous).

What you actually need to do is to curate yourself. For example, if you are a lifestyle blogger, you should try to show viewers your stylish, graceful side. This is the side that viewers would like to see and would like to emulate. Here's a different angle: you're also a lifestyle blogger, but you show your fun and quirky side. This is the cool friend they all want to hang out with, that person who will make them laugh, or that older sibling they wish they had. You might even be someone that they feel like they can confide in. It's still you—but refined and curated.

Love Your Target Market, But Love Yourself Too

When it comes to catering to your target market, I know a lot of you are thinking, "I just want to enjoy myself and do things that I like!" Of course, not everything should be for your viewers, which is why you should try to find the perfect balance between their needs and yours. If you do everything in your power to cater to their wants and needs, what about yours? Have you forgotten about your own interests and your own views?

Once you stop thinking about yourself, your viewers will become consumers of you, and you will lose that human interaction that makes everyone love YouTube creators. This will turn into a regular job that you might soon hate. This whole thing should be something that you're passionate about but just so happens to earn you lots of cash—isn't that why you wanted to be a YouTuber in the first place?

Ask yourself, "Is this what I will find myself watching when I'm on a YouTube binge? Do I find this entertaining? Do I love this?" Create videos that you would find interesting and would spend countless hours watching. When you're ultimately enjoying yourself while doing this, you won't struggle to think of new ideas. If you're enjoying yourself, it will be evident in your work, and most viewers will see that and find it engaging.

So again, find a middle ground—somewhere you would place it in the market but at the same time keep it close to your heart. It's a balancing act, really, the way you would brand your channel. It's better to start with something you love, tweak it, and curate it for your viewers. Then you can use your branding to attract viewers and keep them watching the videos that you love creating with how you talk, how you dress, how you edit your videos, your music choices, and even your scripts.

Take the time to make your scripts

Scripts? You don't want to write scripts, you say? Not good with words? That's totally fine—you just need an outline, then make it up as you go along. Inspiration works like that sometimes. You can just edit it afterward, then reshoot and edit again. Every creative process is different—there is no same path to creativity. Not every video needs scripts, anyway. Of course, if you have a more informative type of channel, you need scripts. But if you do ASMR videos or something along those lines, you can totally get away with no scripts—you just need storyboards. Ha! Kidding aside, all I'm trying to say is you need a plan or a direction to guide you along. I call them scripts for simplicity's sake—some people like to call it storyboarding, but it's more like an outline of what you want to do in your video. It's pretty generic: you start with a beginning, usually your introduction, then proceed with the middle, which is the main content. Then, you come up with the ending, and that's that. Every show you watch is following some sort of outline or script.

Look back and remember all the shows you watched before. Yep, everything has a script—just remember that once you start with one style, you shouldn't diverge from your format too much. And keep your branding in mind! Ask yourself, "Is this me? Is this my channel? Does it fit in with all the other videos?" If it doesn't, and you really want to produce it, then you'd better make another category or a sub-channel. Having an outline as you shoot your videos will also save you a lot of time shuffling about. It ultimately saves a lot of time editing too.

Editing is Your Friend

Speaking of editing, pick an editing style that fits your channel branding. If you're a thrift store fashion channel, the 90's style editing could work for you. Featuring daily life and a bunch of relaxing vlogs? A lot of still shots and slow pans, and scenic montages would work beautifully. Just remember to match this with what your content is. Scenic mon- tags won't work with documentary videos about the current political climate, of course. You will find your editing style with the more videos you edit, so keep at it, and you will get better. Just remember to start with your branding as your base, then slowly move along and improve. It's just like drawing—use the basics like anatomy as your starting point before experimenting with illustration styles. Develop an editing style that is unmistakably yours but also works well with your brand.

Don't Stop the Music

Background music needs to be chosen very carefully because music adds extra oomph to your videos. Even silent films back then weren't really silent—they had a whole orchestra accompaniment with them. So imagine you're a lifestyle blogger who has this graceful, French chic image...and your video montage is set to death metal. That's just too jarring!

Find music that helps with your branding and is totally you at the same time. The music should definitely be something you actually like because, realistically, you'll be hearing it again and again when editing. If your target audiences are tween girls around 12 to 14 years old, you probably don't want to use old jazz music or early 2000's emo.

Trivia: did you know that brands pay huge sums of money just to have the perfect playlist for you to shop in? They only play the music that caters to their market, as it adds this certain feeling that makes you feel happy and energized when shopping. It's the same thing with the scent that you smell when you enter stores. Yes, branding attacks all senses.

That said, you can't just pop in the latest Justin Bieber or Coldplay song and use it in your video. You need to use FREE music or your videos will be taken down by copyright and fast. There are sites out there that offer music that you can use on your videos: Soundcloud, Amazon Music, or Epidemic sounds, to name a few. Most of them have subscription fees, mind you, but some tracks on Soundcloud are free to use. Most of the songs that you can use from here are under Creative Commons. As you can probably sense by now, there are various copyright laws and licensing issues involved with this.

Don't worry; I'm here to help you, right? Here's a quick guide:

First up is "Creative Commons." The main thing about "Creative Commons" is that the musician or sometimes the licensor gets the credit for the music they create. So, music protected under this license usually specifies what you need to do or provide to use their music in your videos. This means that you are free to use the songs, but you need to follow the rules set by the artist.

Then comes "Royalty-free," which a lot of people mistake as free. It's totally NOT. This only means that you pay the artist or the licensor ONLY ONCE for the music you would use, and you can use it as many times as you can possibly want. You wouldn't need to pay the artist or licensor royalties every time you use the song. This applies in stock photos and videos too.

And finally, "Public Domain." This music is totally FREE. If all rights to the music have already expired—it differs per country, so check carefully—then it's on "Public Domain." This means you can do anything and everything with the music under this without paying a single cent.

That said, be careful with your music choices. It's already hard to find the perfect soundtrack to your road trip with friends, so finding music that fits you, is in line with your branding, and won't have you sued is twice as hard. I suggest sticking with free sites first; you can pay for subscription services later when your channel is already up and running. Of course, there is an easier way because YouTube actually has an audio library located within your YouTube Studio. You can even monetize your videos with songs from the YouTube Audio library because even some tracks under "Creative Commons" still specify that you can't monetize them.

In summary, branding is everything, and if you do it right, it can help your channel a lot. Since you're building your channel from the ground up, you absolutely need brand recognition to compete with other existing and new YouTubers. And for you to build brand recognition, you need to find a style or a look that is uniquely you.

Use branding as a means to further connect with your viewers or attract them, because as they say, "People fall in love with people." So, you build your channel using your branding as your guide, but don't forget to keep it tied to your likes and preferences as you are. Think of it this way: you create videos about content that interests you for people who like the same interests. In the end, you would want viewers to go to your channel by default—or, better yet, to actively seek out your channel when they are looking for videos of your specific topic. Again, this is slightly different if you are a business—you also use branding to reach your target market and give that human feel to it. But instead of thinking about the branding and your preferences, businesses think about the branding and the brand itself.

Uploading A Video

And so it begins: you're finally ready for your first upload. The instructions seem never-ending, don't they? Don't fret. It may all look confusing at first, but when you finally get the hang of it, it'll all pay off in the end.

So let's say you have your video file on hand, all ready to be launched. You took pains and made sure it's in the right format, size, and ratio. Sign in to your YouTube account and go to YouTube Creator Studio. On the upper right-hand corner of your screen, you'll see the Create button right beside your profile or channel icon. Clicking this will give you a drop-down menu for uploading your video and live streaming (in here, live streaming is "Go Live"). You can also access these functions by clicking the "Upload" and "Go Live" icons below the "Create" button.

On the Upload page, you can drag-and-drop your selected file, and you will be promptly brought to a page where you can enter the details about your video. Under Details, you can type in the video title, descriptions, and upload your well-crafted video thumbnail and add your video to an existing playlist. You can also create a new playlist on the spot. This page also lets you specify who your audience will be. Is this made for kids? Is there an age restriction? Tags and closed captions or subtitles can also be added to this page.

The Video Elements tab lets you add an "End Card" and various other "Cards." Note: you can't add an "End Card" or "Cards" to videos you marked as made for kids. Adding cards to your video allows you to promote your other videos, other users, or your channel. At the same time, you can also add a Subscribe button and your channel icon here. This could be your CTA or "Call-To-Action" button.

Lastly, under the Visibility tab, you can choose who you can see and when you want to publish your video. The "Schedule" option lets you specify the exact date and time you'd like your video to be made public. This option is a great tool for consistently keeping up with the set schedule for your new video releases. Think of this as the time slot for a series that you are looking forward to (e.g., "new episodes every Friday night!" or something of the like). This can create hype and help make your viewers look forward to your new releases. Made a mistake and clicked "Done" too fast? Don't worry—from your YouTube Creator Studio, and you can access your video by clicking the video icon on the toolbar to your left. Edit different aspects of your videos easily by clicking the detail or pen icon when you hover over the video thumbnail. From here, you can even edit the whole video by using the in-house or in-browser video editor.

The Video Equipment That You Need

When you think of vlogging equipment, your first images are of cameras, microphones, and lights. You may also be scrambling to figure out what is best in each of these categories and how to buy them without breaking the bank.

What if I told you that all that gear isn't necessary when you start out? It's true. If you have a smartphone, you have everything you need to get started.

When you've got an idea and eager to start, this is the best method to use. Don't worry that it might not have all the video options required, though there are apps and accessories that you may want to consider augmenting your phone. We'll talk about those later.

Consider the ease of recording with your phone. It's light and easy to carry and, in fact, you carry it everywhere. When you vlog from your smartphone, you'll never say, "Gee, I wish I could have recorded that moment."

There's another advantage to using this small device. You can skip the step of transferring files to your computer. You can edit and upload them directly to your platform from your phone.

Basic Rules

But best of all, the guidelines for recording from your phone aren't many. Here are the most important four.

Keep the Lens Clean

Our phones take a lot of abuse, so we neglect the need for a clean lens. Any smudge on it can spoil what otherwise would have been a great video shoot. You don't want your message to get lost in a sea of comments and speculations on the smudge.

Keep Your Phone Horizontal

This is essential since YouTube viewers, especially, enlarge the videos and watch them in a horizontal position. Besides, vertically shot videos tend to have an amateurish look to them. Of course, exceptions to this rule exist and, with time, you'll know when a vertical shot feels right.

Steady Your Phone

Sometimes, it's difficult to steady a hand-held phone, even for still photography. It's even more difficult at times in videography. Discover the techniques that work for you to get the most stable shots. It might be two-handed recording, or you may opt for a **gimbal,** a device that stabilizes your phone.

Ensure Sound Quality.

The newer smartphones record quality well. If you're shooting in public, you may be frustrated because it picks up sound indiscriminately. The danger is the background noise will drown you out. Just think of those times you're listening to a live news broadcast at a noisy event. It's difficult to hear what the journalist is saying.

If you plan to record either live events or outside where the noise level is high, consider investing in a Bluetooth microphone or a mini-shotgun one. The shotgun-style microphone is directional: It will be able to highlight the speaker without the background noise.

Smartphone Vlogging Accessories

At some point in your blogging career, you'll want to shoot more complex videos. You may want to shoot moving objects. Perhaps you're at a NASCAR race as part of your travel vlog. Or you may want to invest in lighting equipment to improve the quality of your vlog.

The good news is that you don't have to make the leap to expensive equipment when your budget doesn't let you. You can purchase reasonably priced accessories that work with your phone and still give you the professionalism you're seeking.

Tripod

One of my first investments was a tripod. This is essential for maintaining a steady phone and accomplishing your shoot from different angles. This piece of equipment will pay for itself many times over.

Microphone

You'll also want to think about purchasing a microphone, which will boost the quality of your sound dramatically. The sound on the final cut of your vlog will be clear, strong, and without distortions.

Remote Control

Remote control is an often overlooked addition to your vlogging equipment. It gives you a button to control your camera through Blue-tooth technology. It's another excellent investment.

And for that extra lighting that we all need, you can be creative for the moment. Ordinary household table lamps work well. Play with the placement of these until you achieve the desired effect.

None of these accessories will break the bank, but together they will make a tremendous improvement in the quality of your project.

Editing Videos on Your Phone

Even when it comes to editing your footage before you upload your vlog, your phone is your go-to source - with the help of an app. The good news is, you'll have your choice of several outstanding video-editing apps. And the news only gets better from there because some of them are free.

iMovie

The free iMovie app from Apple is easy enough for beginners to use. It contains the necessary audio tools and more. It also has valuable themes to enhance your episode. One of the better features of the iMovie app is it provides you access to the iCloud drive. An even better feature is the ability to export to YouTube directly. This app also supports 4K video - a high-quality resolution of just over 4000 pixels of width.

ActionDirector

If you have an Android device, you'll be pleased to know the app ActionDirector also has a free version and is user-friendly. It supports a 4K video. With this app, you have access to special effects, such as slow- or fast-motion. It, too, comes with great audio tools. This app allows you to do awesome effects like muting the original sound and adding background music.

One note of caution, though. If you choose to download the free version, you'll get their watermark on your video.

InShot

One of the most versatile free video-editing apps is InShot since it's compatible with Apple and Android. User-friendly, this app's audio tools give you direct access to

SoundCloud, the audio distribution and music sharing platform.

As with ActionDirector, the free version of this app will show a watermark on your video. It also contains in-app advertisements.

Video Equipment Beyond the Smartphone

As you become a more experienced blogger, the odds are you'll outgrow using your smartphone as your camera. When that time comes, certain pieces of equipment will help you create a successful vlog.

Portable Camera

Regardless of the type of vlog you're producing, a portable camera will be the most useful. Beyond that, you'll want to select one with a good resolution, and that performs well in low-light conditions. The other feature you'll want in your camera is video autofocus. While it may sound professional to focus the video yourself, it's a time-consuming activity. Autofocus is the perfect option for vloggers. In fact, I wouldn't buy a camera that didn't have it.

Make sure that among all the features you're considering that this camera has a good quality microphone. It's counterintuitive, but viewers consider the quality of the sound on your vlog even more important than the quality of your video. The ideal microphone needs to reproduce your voice clearly as well as suppress background noise.

The best cameras for vlogging have flip screens, which allow you to view the camera screen during your vlogging session. The result will be a well-focused production with the correct exposure.

Microphone

While cameras have a built-in microphone, you still may want to consider buying a separate mic. As with every other piece of vlogging equipment, you want it to be as lightweight and as easy to carry as possible. If you're vlogging near your computer, you can use a USB microphone. They're inexpensive and provide great sound. If you're in the field shooting your vlog, consider a **shotgun** microphone or a **lavalier** to put on your person.

Lighting

Yes, you've bought a camera that shoots in low light, but you will still find additional lighting equipment useful. This is another investment that pays for itself. By learning how to use lighting, your vlogs will have a professional feel to them. Good lighting can even help improve your scene if your camera isn't that good.

Tripod

If you didn't buy a tripod when using your smartphone for vlogging, now's the time to do it. It's more necessary than ever. Don't

skimp on this just because your camera includes in-body image stabilization. The other advantage is that you can shoot your vlog without the need for an assistant.

Now that you have the equipment to start your recording let's talk about the best way to shoot your video and how to edit it when you're done.

Some Amazing Channel Ideas

Before you decide on your topic and the format, it's always wise to see what kinds of vlogs are trending. Choosing a topic might be a bit trickier than you have originally realized. It's all about finding what I call the "Goldilocks area" of the vlogging industry.

You don't want to select a topic that's overcrowded with channels, which would make it difficult for you to get noticed. But you don't want to choose a subject that is so narrow and unknown that you receive no viewers at all. While the internet can give you a viable audience due to its tremendous audience potential, this is still a valid concern. Your choice should be in that "Goldilocks area" that's not too crowded or too sparse, but just right.

If you know what's trending when you decide your topic, your chances of success will be substantially increased. Here are some of the vlogs people enjoy watching. They cover a broad range of topics and will give you some great guidance on choosing your topic. They're presented below in no particular order.

Unboxing Products

You might already be familiar with this topic. This is a vlog in which a person opens a package, shows the product, and narrates their action through all of it. Children love to watch these when the product is promised to be a toy or game. But your potential audience isn't limited to kids.

You can unbox many types of products, from books to kitchen appliances and everything in between. While the specific video usually showcases the product favorably, it's not designed to be an endorsement of it. Though, as you probably guessed, it comes close to a commercial.

This is the vlog for you if you are informative about your presentation and honest about the product.

Gaming

Vlogs dedicated to video games and game platforms are always popular. These channels focus on all aspects of the games, from gaming to the gaming lifestyle and culture.

There are even channels where you can watch people play.

While this is a popular genre, it's also a very crowded one. Starting one is easy and cheap, though, so if you love everything about video games and can bring something new to the genre, then this is your territory.

Lifestyle

This is another term for the a-day-in-your-life vlog. It's essentially a video diary that you share with your audience. Not surprisingly, it's unscripted - and to be a success - an authentic look into your life. If your life has variety with no two days quite alike, then you may want to consider this.

Health and Fitness

You don't have to be a nutritionist or personal trainer to host one of the vlogs, but it helps. This type of vlog shares workout routines, diets and dieting tips, and even information on vitamins. What you talk about is up to you.

It's not uncommon for a vlog to be shot at a gym while the subject demonstrates how to use weight machines.

If you are interested in this topic, you should look into it further, especially if you have overcome a physical difficulty or a disease like cancer and took steps to improve your health as a result. Others can benefit from your story.

Beauty

This vlog concentrates on beauty, fashion, and related topics. This genre has a casual feel to it. Usually, it uses the "talking head" format for one person creating an impression they're talking to you. If this is your area of expertise, you're in luck. Of all the different types of vlogs, this is one of the simplest to start since the equipment you need is minimal. You can confidently start with nothing more than a smartphone or a computer with a webcam.

Information

How many times have you turned to a video to learn how to do something new? One of my hobbies is crocheting, and I rely on vlogs for all aspects of this activity, from learning a new stitch to tips on color choices.

Vlogs that inform are always popular and in need. And if you have an essential skill or knowledge people want to learn, then you've got yourself a successful enterprise. Viewers tend to watch these videos more than once, and each time they do, it's credited as a separate view, which helps your rankings.

Reviews

Are you an avid movie-goer? Or are you an unabashed television viewer? You might even be an unquenchable reader.

If any of these describes you, you may want to consider a vlog devoted to reviewing your medium. Many individuals prefer to get an idea of what the story is about and whether it's worth their time before investing in it. If you've never searched for a review on YouTube, try it. People want reviews. If you can interestingly provide insight on the medium, you may have found your niche.

Travel

If you already travel for work or for pleasure, this type of vlog would fit in well with your lifestyle. Who doesn't love to see faraway lands? Or even locations relatively close that they've never visited. You don't necessarily have to travel to Italy for your vlog to be successful. You may want to highlight the interesting and historical locations of a region or your state.

Entertainment

There certainly is no lack of viewers for channels devoted to entertainment, pop culture, and celebrity gossip. You don't need to have the inside track on this type of news - though that would be awesome - you just need to have an opinion. If you're already following entertainment news closely, this might be natural for you. As with all the other most-watched channels, you'll have to bring your unique insight to your topic.

Motivational

These vlogs are in high demand now because we all could use words of encouragement. From business-inspired topics to spiritually uplifting messages, there's a need for all types of motivation.

Do people consult you when they need words of wisdom and hope? Do you have a personal secret to happiness or success? This vlog genre is yours to claim.

Comedy

If you've ever dreamed of being a stand-up comic or if you keep your friends and family in stitches, you should consider this genre. You can find comedy videos that range from comedic monologues to original skits populated with creative characters.

Technology

If your area of expertise doesn't lie in comedy or motivating others, perhaps it lies in understanding the ever-rapidly expanding world of technology. It's not an exaggeration to say new technology bombards us every day, and many of us are hopelessly mired in it.

That's where your tech vlog comes in. Your topics could range from resetting a smartphone to fixing a computer glitch. There are plenty of people out there struggling with these issues. More than once, I've turned to videos to help me figure out how to deal with a stubborn laptop.

This genre can take at least two forms. The first, which I've referred to, is the how-to videos. The second is the setup videos in which you walk your viewers by setting their new equipment.

But it doesn't have to end there. You can show viewers how to make money using the technology they own. If you're savvy with social media, you may even help people generate a larger following.

ASMR

I would be remiss if I didn't mention one of the fastest-growing genres on YouTube: **ASMR.**

The initials stand for Autonomous Sensory Meridian Response. It's the name coined for any physical sensation that soothes you like someone gently brushing their fingertips over your skin.

The topics in this genre vary widely. But they nearly all are the same in one respect: The message is delivered in a whisper. There are also episodes in which only relaxing sounds are played. One of these episodes is nearly two hours of a working hairdryer.

This is not your traditional vlog topic. But, the two most popular vlogs, Gentle Whispering ASMR and ASMR Darling, boast more than one million subscribers. Not only that, but they've also partnered with brand names. When you've made that decision on your topic and format, it's time for the step you've been waiting for: creating your brand. Read on to the next chapter to find what's involved

The Advantages And Disadvantages To Using YouTube

Free To Use

Before anything else, the most basic and one of the most important benefits of YouTube that we all have to appreciate is free to use.

Unlike other apps that collect payment before you can access or use them, YouTube is free of costs - from signing up to watching videos; you won't have to pay for anything.

As we all know, YouTube is a famous platform for earning money, and with that being said, it is really an advantage for you to be able to use all their features without risking some cash.

Watching and uploading videos is completely free, which allows you to maximize all the available features and test out whether the app is suitable for you or not without worrying about the charges.

However, YouTube has recently offered a new paid subscription called. YouTube Premium allows you to access special features like watching videos without ads, playing games offline, playing videos in the background, and many more. In addition, this new feature also allows you to access YouTube Music Premium, where you can watch YouTube original movies and series.

Regardless, whether you avail of the free YouTube feature or the YouTube Premium, you can still upload videos and access YouTube's basic features, which are important for you to start making money.

Convenient

YouTube can be used anywhere and anytime on various devices as long as you have a stable internet connection and a YouTube account.

It is very convenient to use because you won't need to download it. Most of the time, since YouTube is owned by Google, you will find it automatically installed on your mobile device or if you opt to use it on your laptops and computers, you will just need to open your browser.

Aside from the convenience it gives, it also serves as a medium to make money from home. It is one of the best online platforms today to make money online using your resources and means.

Great Useful Content

Suppose you aren't a celebrity, or you aren't eye-catching enough to gain a million followers. In that case, one of the most common reasons why your audience still sticks to your channel is because you are giving them great and useful content that can help solve their problems.

When you have a YouTube channel, you can make a hobby out of it, but you can also share valuable information and content with your audience, allowing them to stay longer and visit your channel from time to time.

Having great content on your YouTube channel is like having a best seller book that readers come back to every so often whenever they want to feed their minds with information. Even though there are many books out there, they would still want the same book from the same author.

Thus, in YouTube, having at least one good YouTube content can make you closer to having a thousand followers and establishing a spot in the industry.

Personal Connection with the Audience

YouTube is all about connections, and if you aren't able to establish a connection between you and your audience, you would not most likely make it in the industry for long.

Having a close connection with your audience gives them a sense of familiarity, which is an important factor in building a community. Significantly, YouTube communities affect the success of a YouTuber.

Many YouTubers often use their connection with their audience to gain a stable amount of views and subscribers. This is relevant because the audience can relate to you by watching your content.

Having a YouTube channel allows you to introduce yourself so that they can get to know you more and your content.

Creative Outlet

Undeniably, YouTube helped many artists, professionals, and ordinary people showcase their talents and skills without exerting a lot of time, effort, and money. With the help of a stable internet connection, a laptop or cellphone, and a few skills, you can now reach other people all over the world.

YouTube has been a creative outlet for many individuals to get known and get paid while enjoying their craft. Having a YouTube channel also boosted this generation's creativity when it comes to presenting themselves in front of the camera and editing their videos.

Innovation

Today, it is without a doubt that technology has dominated almost everything in the world. Suppose you want to collect information, learn something, or just communicate with someone. In that case, all you need is to create a social media account to gain access to everything.

YouTube has provided a way for people to start modernizing and learn the benefits of different innovative techniques. Since producing videos are viral and trending nowadays, video influencers found a good way to stand out in their approach.

Source of Traffic

One benefit of having a YouTube channel is that you can drive additional sources of traffic to your website. For instance, you have a local business or promote your website or social media accounts. Having a YouTube channel to extend your content is an effective technique to gain a larger audience.

If other people are not knowledgeable of your site or account, using YouTube as a platform can be helpful. Instead of posting photos or blogs alone, you may opt to create a YouTube account and share some short clips or commercials regarding your products and videos.

Additional Source of Income

Ever since YouTube has become a trend, many people, including celebrities, models, artists, and ordinary citizens, shifted from their professions to being vloggers because of the good revenue.

Indeed, YouTube is a good additional source of income. Whether you are earning from Google ads or through promotion, it creates more traffic sources to your channel, allowing you to gain an additional source of income.

Even though you don't regularly upload, as long as there are still people watching your previous videos, your revenue will continue to grow.

As much as there are numerous reasons why YouTube is indeed beneficial for many people, there are also several downsides to having one.

1. Everything is public.

YouTube is a social media platform which means that everyone who has access to YouTube can see everything that goes on the app, including the one that you posted a few minutes ago.

Once you have uploaded a video publicly, there's no turning back. This means that whatever you have posted on YouTube is not only yours anymore. Hence, everyone who watches or sees it has full control of it.

They can download, edit, or reproduce it the way they want because the content was already shared with the public.

While it may sound like a not-so-serious problem, it can cause privacy invasion for other YouTubers.

Nowadays where everyone comes up with an issue about almost anything, it is not impossible for anyone to be a victim of cyberbullying and harassment.

You need to understand that the moment you decide to share your personal information or your own opinion of a certain topic in public, your privacy and reputation can be majorly affected.

In the history of YouTube, there were already countless controversies and issues that involved the personal life of many celebrities and vloggers. This is because their privacy was put at stake, given the fame they enjoy.

However, today, most vloggers have turned this downside into an opportunity for them to attract a larger audience. They use these issues and scandals as clickbait.

Whether you become a YouTube sensation or not, it only means one thing - you can be the center of attention either positively or negatively. Thus, you should be prepared to face whatever consequences might get in your way.

Terms and Conditions

Like any other application, YouTube has its own set of terms and conditions that can limit your activities. Since you will be using YouTube's features to earn money, you must ensure that you comply with all their rules to keep your channel working.

YouTube has many rules, especially today that several users have been using it. The terms have increased, and they became more complicated.

Suppose you have been a creator on YouTube for around some time. In that case, you will notice that YouTube changes its terms almost yearly, which is a disadvantage, especially for newbies (because they have to deal with all of the changes before they get monetized).

Third-Party Company Can Interfere

Once your channel has passed all YouTube's requirements, you will start running ads on your videos. Thus, this allows advertisers outside of YouTube to use your videos to endorse their brand and products.

This would also mean that your competitors can also run ads on your videos, affecting your brand.

Your Account Can be Disabled without Notice

YouTube can disable your account if proven that you have gone against the community standards and guidelines. This can happen if you have done the following:

The content that you have published is owned by another content creator. This means that the owner has obtained copyright of the video, and you have no right to use it under your usage.

The content you have published contains violence, nudity, harassment, spam, etc., which are all a breach of YouTube's terms and conditions.

You used your content to deliver false information, hate speech, etc., which may ruin another person's identity or lead to an immoral influence on viewers.

Your video uploads have been repeatedly reported for copyright infringement claims.

Aside from these, many other factors could result in your account getting disabled, suspended, or blocked. At times, a mistake can even mess up your account.

For instance, you have commented multiple links on the comment section of a YouTube channel that you are following. The next day, you found your account already restricted, which means that you cannot access it anymore.

This could occur even though the links you have attached do not go against YouTube's community standards. What is more, you won't receive any notifications about your account getting terminated.

Dealing with Negative Comments and Trolls

As much as YouTube provides a great room for opportunities and engagement to you and your audience, it can also create an entry for haters and negative commenters to pool over your videos.

There can also be "trolls" invading your channel and videos that can throw undesirable comments and ruin your brand's reputation. Although this might be uncontrollable to some extent, YouTube has provided feature *Comment moderation guidelines* where you can be able to turn off comments even though your video is public. To do this:

Sign in on your YouTube account.

Go to YouTube Studio. In the Settings tab, choose the *Community* option

Under *the Community* tab, choose *Defaults,* then click the options under the *Comments on your Channel.*

Here, you have the option to allow all comments, hold other comments for review, or disable anyone from commenting.

If you wish to disable all comments regardless of what kind, choose the *Disable Comments* option, then Save.

By using this feature of YouTube, you get to minimize the visibility of comments on your videos. Although you cannot control what others think or say about your content, you can avoid spams and trolls from loitering in your channel.

Monetizing Ideas For Your Channel

Earning money from YouTube is simple. As long as you can upload videos with great original content, and they were able to gain a good amount of views, you are sure to make money from them.

As this is the usual case, the YouTube program has extended its features to provide YouTubers with the opportunity to earn more than just uploading content alone.

Today, aside from earning a share from the advertising revenue, YouTubers are given the access to monetize their content through company partnerships, sponsorships, endorsements, crowdfunding, and product merchandising.

YouTube has undeniably given a vast opportunity to YouTubers in terms of making money from uploading multiple videos and garnering millions of views to being one of the high-paid top influencers on the Internet.

Today, YouTube has been a popular tool that many people use to earn either an additional or steady income. With the many opportunities and advantages YouTube has, several self-made celebrities have earned and made a name in the industry without making videos.

For many YouTubers, producing creative content regularly may be difficult. This is because of the guidelines, terms, and conditions that must be strictly followed, or it could also be for other reasons like time management, creativity, equipment, or editing skills.

Suppose you have already made a YouTube channel. In that case, you know that before you can monetize it, you must first acquire at least 4,000 watch hours and a minimum of 1,000 followers. Aside from these, your channel will be checked afterward to ensure that it does not go against the community standards. Terms and conditions are applied, and each one must be strictly observed.

After this initial process, YouTube content creators are given access to earn from Google advertisements, fan funding, or selling of merchandise. This opportunity has made a lot of individuals become millionaires and instant celebrities.

Become a YouTube Partner

Before your YouTube account can run ads and get monetized, you must first join the YouTube Partner Program. The YouTube Partner Program is a YouTube feature where content creators are provided with access to the Creator Support team and other direct YouTube resources.

Joining the YouTube Partner Program does not require you to get a partner. You only need to set up your channel, connect a Google AdSense account, and get a good amount of views to get going.

The benefit of being a YouTube partner is you have access to multiple income streams. Aside from earning from ads, you also have the right to YouTube premium subscription, channel memberships, and access to sell your merchandise.

Whether you want to earn money by producing videos as a content creator or you intend to search for companies to hire you for promotion, the first way to start earning is by applying for the YouTube Partner Program.

To join the program, you must first meet the requirements. Currently, YouTube has set a new policy to get monetized. You must first acquire at least 1,000 subscribers and 4,000 public watch hours over the past year. Once this is done, you are now granted many cash-in options.

Nevertheless, your content must pass YouTube's guidelines and policies. You should also have a Google AdSense account linked to your channel prepared for verification.

Brand Sponsorship

Brand sponsorship is another known source of profit aside from running ads on your videos. It is also known in digital marketing as "influencer marketing" because companies mainly work with various content creators and influencers to promote their products and services.

Suppose you do not want to be a full-time YouTuber to earn money or find it hard to create video content regularly. In that case, you can still earn money on YouTube by being a brand endorser.

One great example is when Michelle Phan eventually became a brand endorser of a French luxury cosmetics company. Michelle Phan is a famous make-up artist and YouTuber who started her YouTube career by creating make-up tutorials.

Since she was one of the pioneer makeup enthusiasts on YouTube who got famous, various cosmetics companies started getting her for endorsements and promotions. One of these is Lancome Paris, a French luxury perfumes and cosmetics company under the L'Oreal Luxury Products division.

Aside from doing YouTube videos, Michelle Phan has undeniably earned a huge income from product endorsements and brand sponsorships.

Currently, almost all companies and agencies sought YouTube personalities and influencers to endorse their products and services. They also look up other social media platforms like Facebook, Twitter, Snap-chat, and Instagram.

Crowdfunding or Fan Funding

Crowdfunding or fan funding is one of the most common ways individuals or organizations use to collect money. Basically, top social media platforms like Facebook, Instagram, and YouTube are used to collect donations from family, friends, friends of friends, and strangers.

Being one of the top social media platforms today, YouTube is highly influential and quicker to spread awareness than traditional methods. YouTube has billions of viewers, and it is not impossible to attract many potential donors by using this platform.

Fan funding is also a good way to interact with your fans or other colleagues. Thus, you can better present yourself, your ideas, and your cause for earning such an amount by using this opportunity. This way, you may even acquire more fans and donations in your succeeding plans.

If you choose this method, you should ensure that you follow a couple of practices to avoid confusion and misinterpretation of your project.

Your priority is to focus on the project's cause. Create transparency on how you would use the money. This will allow your fans and followers to understand your story and get invested in your mission. Once they do, they will buy everything you would post because you have already gotten their loyalties.

Second, you would want your fans to know that they matter and that you are gratified with their support. You may offer enticing prizes and rewards to selected fans of yours not only to extend appreciation but also to gain more exposure from fans and other people. Most likely, your followers will monitor your every post just to acquire that rewards. Hence, the more you practice this, the more likely you will have more pledges and donations over time.

Sales and Merchandise

There are many products that you can sell with the help of your YouTube channel. Whether you are selling your merchandise or you were invited to promote other people's products, you can get a fair share of revenue through this method.

YouTube partners have access to sell up to 12 items on YouTube. This feature is available in the accounts of YouTube partner program members who have acquired more than 10,000 subscribers. Two best examples of

YouTube stars who have made millions from YouTube merchandise are Jeffree Star and Drake.

Jeffree Star has not only been known for producing cosmetic and fashion videos but also for selling lipstick, eye shadow palettes, and highlighters to her fans. Even Drake, who is already a successful celebrity rapper, has undeniably earned millions from selling shirts.

Meanwhile, for example, you have acquired thousands of subscribers for a video you have previously made. You have been inactive for quite a while. You can still be granted this feature, provided you are eligible.

If you are that person who isn't good at being creative in terms of producing videos regularly, but you are good at handling business, you can be able on YouTube by selling.

Alternately, if you have been an entrepreneur first and then a content creator second, you can still earn by designing your own YouTube marketing strategy, given the eligibility requirements stated above.

Anyhow, merchandise is a worthwhile method to make money from YouTube.

Re-Uploading

Re-uploading is another viable method on how to make money on YouTube without making videos. Although this can sound easy, re-uploading videos may require a lot of effort and creativity.

When you re-upload other channel's videos, you don't want to upload them that way without editing.

For your channel to earn, you must know how to create your own identity and trademark. Though, YouTube allows reposting and re-uploading of videos given that they fall under the Creative Common License. Thus, all common creative videos are eligible for monetization because they are set up for free distribution and reuse.

Chapter 6

Earn Money Through Freelancing

Working as a freelancer means running your own business rather than being employed by someone else. Freelancers are self-employed individuals who are also known as independent contractors.

Freelancers are hired on a part-time or short-term basis by other companies, but they do not receive the same pay as full-time employees or have the same level of commitment to any particular company.

People are talking about freelancing more than ever before, thanks to the rise of the gig economy. And this is because there are more freelancers today than ever before – according to a 2019 study by Upwork and Freelancers Union, 57 million Americans freelanced in 2019.

According to the same study, 53 percent of workers between the ages of 18 and 22 freelance.

When it comes to working that a freelancer may do, not everyone uses the terms "freelance" or "freelancer." Indeed, the majority of freelancers refer to themselves as "self-employed."

Deciding What Service To Provide

These days, almost every service can be outsourced. Businesses are learning that they don't need to employ someone to sit in an office to answer phones or a web designer to work on their website. Now they can just hire someone at a project-based rate. When they're not needed, then they don't need to pay them. This is true even for something traditional such as receptionists who answer your calls. You can set up anyone anywhere in the world to answer your phones, take messages, make appointments, and manage the general day-to-day operations of your business. The Internet and technology are opening up many opportunities for people to step in and make a good living, all from the comfort of their own homes or anywhere in the world. You could be relaxing on a tropical island and still be making some significant money while you drink cocktails and soak up some sunshine.

You will need to get yourself set up to start freelancing, but this doesn't have to be expensive. These days, you can purchase yourself a reliable laptop for anywhere from

$30041,000 depending on how much money you are looking to spend. If you have a laptop or computer, then you are halfway there already. Just make sure that it's reliable. Another thing to consider when you're trying to decide between a laptop and a regular desktop computer is whether or not you'll be traveling. A laptop is a lot more convenient if you plan to work away from home, even if you plan to move around inside your home. You'll also need to make sure that you have a reliable Internet connection. A decent home connection is perfect, and you can even get a mobile internet USB stick so that you can take your Internet with you if you're traveling. You may need to purchase some new software programs so that you're up to date with the latest editions. Don't rush out and buy everything to begin with; just purchase what you need to have initially.

Set up a PayPal account or bank account, and you're ready to get started!

So now that you are prepared to step into the exciting world of freelancing, you have some things that you'll need to work out. When you set out down the path of freelancing, the first thing you'll need to establish is what service you will provide. You have to be realistic here, and there are already a lot of this, so don't expect to get jobs as a graphic design artist or engineer unless you actually are one. Suppose you are qualified and competent in whatever area you choose to focus on. In that case, you will soon start getting clients and moving towards the next steps of your freelancing career.

Just because you don't have skills such as a website designer or publisher, it won't eliminate you from becoming a freelancer. You may have other secondary skills such as typing, answering phone calls, data entry, etc. The list of skills that people are looking for is endless. You don't have to be able to do everything that an office administrator does to succeed as a freelancer. That's the great thing about it. People are looking for a variety of different freelancers who can do various aspects of the work they want. Sure, some people will be looking for someone qualified to do everything, but you will soon find out that you can carve out a niche market of work for yourself.

Popular Freelancing Services

Here is a list of some of the most common freelancing services that people are supplying:

***Marketing* -** This can be broken down into a lot of smaller categories. Marketing Coordinator, Project Manager, Marketing Assistant, Marketing Manager, and Research Assistant.

Business Project Management - There are several smaller branches to business management. Project Manager, Project Analyst, and Project Manager's Assistant.

Web Development - This has several branches that you can choose from. Web Application Developer, Junior Developer, Web Content Developer, and Web Development Manager.

Writing - With a little skill and creativity, you can definitely get a great start in writing. Content Manager, Ghost Writing, Web Content Writing, Editor, Technical Writing, Blog & Article Writing, or Proofreading and Review Writing.

Accounting - If you have any skills with numbers, this could be for you. Accounting Clerk, Property Accountant, Accountant, Bookkeeper, and Financial Consultant.

Tutoring & Teaching - There is a world full of people looking for language instructions. Teacher, Tutor, Language Tutor, Teacher's Assistant, and Research Assistant.

Social Media - This is a broad group. Pinterest, Facebook, Twitter, Email Campaigns and management, Advertising, Forum Management, and Website Management.

Graphic Design - Digital Designer, Visual Designer, Graphic Design Assistant, Communications Designer, Website Design, and Graphic Artist.

Administrative Assistance - Here is a good place for beginners to start in the freelancing industry. Executive Assistant, Personal Assistant, Virtual Assistant, and Administrative Assistant.

Even if you only have core competencies in any of these fields, then you should be able to work yourself in at an entry level. Start by picking a general area, and then work out what you can and can't do. There are also many courses available online or at community or technical colleges that will help get your skill levels up.

When you're applying for positions as freelancers, people will be less interested in seeing a resume as such and more interested in seeing results. If you do a small job to prove that you are capable, that will be enough to secure you a starting position in most cases. Remember, good results and prompt, clear communication are among the most desirable traits that employees look for in freelancers. They want to know what you're doing when you will be finished and that you understand what they require from you.

Just be sure to do your research before you start out freelancing. Common sense is the golden rule when considering any business opportunity, no matter what it is. The world is full of people who are looking to take advantage of others. Remember, if it sounds too good to be true, then it most likely is. Do your homework, read online reviews, and in most cases, you don't need to hand over any money to start on freelancing sites.

The Importance Of Branding Yourself

Once you have decided which service you will provide, it is time to start building up your online reputation. Before you can just get yourself a website and have clients looking for you, you'll need to build your reputation and brand. Freelancing can be an excellent way to build client lists and have a base income coming in before you branch out on your own. You may earn less than usual freelancing initially, but the benefits, in the end, could be well worth the reduced rate in the beginning. Consider this as a promotional discount on your brand, much like you would have a discount on a product you were considering releasing.

Freelancing online is not the same as applying for jobs in your everyday life. Potential employers are less interested in what you have done earlier in your career and more interested in what you can do for them and their business.

In theory, you could have been fired from several jobs or have an adverse recommendation from an employer, and they won't care. Potential clients are more interested in getting an excellent product at a good price; actions speak a lot louder than words in the freelancing world. When you start freelancing online, it is, in essence, a new beginning, but this doesn't mean that you can afford to throw away opportunities.

There are a lot of different freelancing platforms available online for you to choose from. Many have similar formats, and some may require you to pay a small fee for you to bid on jobs that become available. Others require you to pay to boost your applications, and the rest are entirely free. What I recommend is that you choose several of the freelancing sites, build your profile, and then drop them as you see results or not. This way, you will have several different options on the go, and you can easily work on more than one platform at once. If a site requires little attention and you only receive one or two good jobs, then it doesn't cost you much to keep that site active to attract potential clients. You never know when the next small client turns into that great client that keeps you busy for years down the track.

The first thing that you will need to do is log in and create an account and user profile. Many sites will allow you to upload a video to your profile, which is always recommended to boost potential views of your profile. You will need to link your profile to a payment method also, such as a bank account or PayPal account, and some sites will require this name to match, so remember that also. When you create your profile, choose a real picture of yourself to make your profile seem as realistic as possible. The more people feel as if you are a real person, the more they will feel comfortable interacting with you.

Now that you have your profile created and set up your account, it's time to move onto the next stage. Before you start applying for jobs and projects, there will be, in most cases, tests that you can do to prove that you have some core competencies in whatever field you have chosen as your service. These tests will either be free, or you may have to pay a small fee for every test that you wish to take. They range from basic English language skills, email skills, social media tests all the way through to programming and graphic design work. Showing that you are already somewhat qualified can be a big advantage when you're applying for jobs. If the tests are free, I recommend doing as many as possible if you are experienced in them. Building up your profile as much as possible will help you when applying for potential jobs, but and are looking for freelancers.

The way freelancing sites will vary from site to site, no two are exactly the same, but all are similar enough that you should be able to create a profile and find your way around. Getting those first few projects will always be the hardest obstacle to overcome. Once you have completed some jobs and received good feedback from clients, every job after that will become easier to land.

The way freelancing sites work is similar in most cases; the fundamental idea is that you create a profile and either search for work on the site or let employers find you.

Let's take Fiverr, for example. The basic idea behind Fiverr is that your gig or job is five dollars. You can then add extras to build up your position to the amount you want. It's an excellent platform to get started on, as it will cost you nothing to set up and start working. Fiverr is a lot more buyer proactive than other freelancing sites. Customers are more likely to come to you looking for your services, but there is also a small buyer request area updated daily that freelancers can apply to. Having a good profile and feedback will make your profile more likely to pop up for potential clients.

When you go to a freelancing site such as Upwork (which is a merger between oDesk and Elance), clients will post their projects, and it is up to you to apply for these positions. Same thing here with your feedback and profile, the more complete it is, the more attractive your proposal will be to potential clients. There will be a daily, weekly, or monthly limit to how many positions you can apply to with most freelancing sites. This is to prevent people from applying for every position and ruining the system with bad applications. Some freelancing sites may make you pay a small amount to apply for jobs, or they may make you pay if you want to go over your limit.

It's strongly suggested that in the beginning, you offer services as cheaply as possible until you start to either build up a list of clients or get yourself some good feedback. There will always be plenty of people looking for the cheapest work possible in the quickest turnaround time in exchange for good feedback. Once you get some work coming in, and you begin to have clients coming to you, then you can start to choose

what work you want and what you will charge. Some clients are happy to pay more once they have formed a relationship with a good freelancer; others will move on to the next cheapest freelancer that they can find. Try not to consider your lower prices as a big loss too much at the beginning. Instead, consider this as promotional work to build your brand and profile on the freelancing platform.

There are a lot of different freelancing platforms available online. I'm not going to list every single one out there, but

I will list the top 25 freelancing sites for you to get a start on. Remember before you hand over any money for anything; if it sounds too good to be true, then it very well could be. Take your time, and do your homework on the freelancing sites, read the reviews online, and consider them before you invest any money into them.

Trusted and Popular List of Freelancing Sites

These first sites are freelance sites more suited for content writers, data entry operators, website designers, developers, programmers, and graphic artists. These freelancing sites will also offer plenty of opportunities for virtual assistants and any other online job position available.

www.freelancer.com www.upwork.com

www.guru.com

www.jobs.problogger.net/

www.craiglist.com

www.ifreelance.com

www.peopleperhour.com

www.gofreelance.com

www.demandstudios.com

www.project4hire.com

www.logomyway.com

www.fiverr.com

www.programmermeetdesigner.com

www.allfreelancework.com

www.projectspring.com

www.online-writing-jobs.com/

www.projects.co.in

www.journalismjobs.com

www.freelancewritinggigs.com

www.ejobs4pros.com

what you decide these are; your brand needs to reflect on your ideas of what makes you unique. You need to think about what you are going to offer and how people will look for your services.

If you provide more than one service, pick which one you will focus on, and then this will need to be reflected in your branding. There aren't many points trying to brand yourself as a "freelance writer, photographer, website builder, and graphic design artist." It's just too much going on, and you will only get a small return on each instead of a real return on one. You can still provide all of these services, but try to stick to one or two at most for your branding.

Do Your Homework on Your Business Brand Name

Now that you have determined which service you will focus on, then it is time to start thinking about whether or not you're going with the catchy business name or using your name as your branding.

This is where it will pay to do some research before you commit to anything. Do some research online, check your keywords, and see what else is out there online. Check available website addresses and prices to obtain them. Some are quite expensive, while others will turn out to be less popular but more affordable.

It is Time to Choose the Appropriate URL For You

Your name as your URL will always be the cheaper option, to begin with, unless your name happens to be the same as a famous movie star, in which case you could be in trouble! Your branding business name will be more expensive, depending on what services you are offering as a freelancer. Whichever way you go, it's a good idea to buy your name in either way. This way, you have locked it down for later if you become famous! You never know when it could become useful to you in the future.

When you choose your URL, try to get some great search words into your URL. Try to think about what people will write when

searching, what words they will combine, or what services will most commonly be referred to. You always need to think about how people will write when searching, not just what services you offer as a freelancer.

Remember to keep your branding consistent here. Use the same keywords as you have in your URL across your website and social media campaigns.

Choose Yourself a Catchy Tagline

Try to keep your tagline as short as possible. A full paragraph isn't going to work. Keep it simple and to the point so that people will be able to not only remember it, but it will be more likely to pop up when they search for it.

Now it is Time to Think About Integrating Your Brand into the Website Design

Remember that branding isn't just the words that you choose as your tagline or the way you describe yourself or your services. Your branding has to include not only words but also images and your style. When you create your website, remember that if you have a particular color, style, logo, or image, then continue that through your website's theme and any email correspondence or when you deliver your work.

Try to keep in mind your name and your business brand when choosing a logo or style. If you can think of a cool or catchy play on your brand, then now is the time to go to a graphic designer freelancer and have them draw up something for you.

A great photo of yourself looking professional, confident, and happy will go a long way towards making people have a much better connection with you as a person.

Developing Your Freelancer Bio

People want to interact with you, and they need to feel as if they are talking to a real person. This can go a long way towards people coming back to you and feeling like they have established a professional and personal connection with you both as a freelancer and a person.

A short bio that leads into your About Us page is a good way for people to feel more of a connection with you.

Now You Need to Reinforce Your Brand on Your About Us Page

This is the second most visited page on any website, regardless of what it is offering. This is your chance to tell your story and sell yourself to potential customers. You want to not only tell your story, but you also want to sell yourself and sell your services to potential clients. Let them know what you do and how you can help them solve their problems.

If you are not a writer, use your Freelancer account to find a freelancer to help you write your website content for you. This can also be a useful way to get to know the freelance sites you are working on. Establishing relationships with other freelancers can also lead towards collaboration or recommendations for future work. They might need your services and recommend you to their clients, and vice versa.

Tell Your Story with Your Online Resume

This step isn't compulsory, but if you feel as if you need to include it or have a highly impressive work history, you may want to include it. Remember, though; resumes can be extremely dry reading, even in some cases verging on plain boring, so make sure that you keep it as light as possible.

Stick to the great work you have done, and include pictures and visual displays or examples of your work. Don't list every job you have done. Consider this resume as a highlight reel of only the great work you have done.

Add Helpful and Interesting Content to Your Website Through Your Blog

This can be a great way to draw people to your website and keep them coming back. If they are interested in what you are posting on your blog, they will need to check back in. This will keep you right in their perspective when it comes to hiring a freelancer.

Even if you're not a writer, this is another great opportunity to get involved with other freelancers and build relationships. Consider swapping services with someone. If you build websites and write, you could help each other out and save money.

This is a great opportunity for you and them to cross advertise on each other's sites. Include a small link, advertising who writes your content or builds your website, for example.

Compose a Quick and to the Point Brand Speech

Keep it simple, and let people know how you can help them. Don't have a long list of what you do, how you'll help, and what the results will be. Keep this short and sweet, and you will find yourself converting more leads.

Keep Your Brand Flowing Through Your Social Media Accounts

If you utilize Twitter, Facebook, Pinterest, Tumbler, or any other social media platforms, remember to keep your freelancing brand flowing through all of the same. If someone sees your Tweets, then you want them to be instantly reminded of your website.

Have your website logos and brand logos designed specifically for all of your social media platforms and accounts. Have a professional create your banners, logos, and profile pictures so that they flow well across your entire brand. A few dollars is a wise investment on a quality image or banner compared to something you try to come up with yourself.

By creating a strong brand presence across every aspect of your freelancing business, it will help people create a link with your brand. Once you have a successful brand, it also opens you up to sell your branding services to other freelancers just starting out in the business.

Good luck with your branding, and remember, keep it clear and keep it consistent, and the results will soon start to speak for themselves. Build your relationships with other freelancers that will eventually start to lead towards mutual work and shared clients. You never know which client will be the one who opens you up to a long list of future clients and a potential lifetime of opportunities and employment.

Expanding Out

We have covered some aspects of your website and your branding in the previous chapter, so I will try not to cover too much of the same information again.

When you first dive into the online world of freelancing and working, in general, you will come to a point where you feel that you need your own website. If you have never built a website yourself and don't have much knowledge in this area, then the task can look simple but turn into a nightmare rather quickly.

There are a few different ways that you can approach this:

- Build your website using a website builder.
- Build a website from scratch by yourself.
- Pay someone to build your website for you.

All of these are good and legitimate ways to approach the problem of creating or obtaining your first website. They will all vary in price, from free, to minimal, all the way into quite an expensive venture. It pays to do your research.

Building a Website Using a Website Builder

We will start with building your website using a third-party website provider. The following sites will be a good place to start:

Blogger

WordPress.com

Tumblr

TypePad

Wix Etc.

There are hundreds of these sites available online, and they all offer services ranging from free to monthly or yearly plans. There are a lot more sites out there, so I'm not going to list them all as they could fill an entire book! Do your homework, and choose one that offers a plan to suit your needs. Most of these sites offer basic packages for free, but you will need to pay for premium packages if you want more services. For the most part, they include hosting and creating a domain name as part of their package.

For someone just getting into working online and creating a website, this could be a quite attractive way to get a basic website at little to no cost. They offer users the ability to drag and drop content and images into selected themes and templates. If it is something more advanced than you are looking for, you may want to build your website.

Building Your Website

Suppose you are Internet savvy already and consider yourself quite capable of learning some basic website-building skills. In that case, you may want to consider building your website. You'll need to purchase a domain name and URL and organize hosting yourself, but this is a relatively simple thing to accomplish. Depending on what, you can be quite inexpensive. Do your homework and shop around when you are looking at hosting options. Look at their speed of hosting offered and the support service that the company offers. Check online for any reviews of the company, and compare it to the opposition. Research with anything you do online is a crucial part of any project. A little bit of time researching can save you a lot of money down the road and many headaches.

This is where we discussed choosing your name in the previous chapter about branding. When you have your name chosen, you can look at purchasing it and getting ready to start building your website. WordPress.org, not to be confused with WordPress.com, is a simple and easy way to start building your website. It allows you to choose from thousands of themes and also to create and edit your own. It may seem complicated at first, but they offer you a lot of support, and many tutorials are available online to help.

If you need help with website content, images, or graphics, you can choose to go back to your freelancing sites and hire freelancers to help you. They can come up with the different parts of your website that you are struggling with.

This is only a quick guide on the key points; I could write an entire book on building a website. You will need to tweak and change many different things as you go along, but by this point, you should have a basic website with some content and images.

At this point, you will be more confident with your site and starting to look into things like SEO and the way the site functions. You can back up your site as you work on it and then restore it to previous versions if things go horribly wrong. Going online to find the answer to any really difficult problems you may encounter is a quick and cheap way to fix anything.

No matter how many websites you build, you'll always end up with small problems or the need to learn something new. The more you work on your website, the easier it will become and the quicker you will get at it.

Paying Someone to Build Your Website For You

If you have no interest, time, or skill in building your website, you can choose to have one built for you. This is where having connections on your freelancing platform can really pay off. If you know someone who builds websites, then consider approaching them to work out a mutual work exchange. If not, you will need to work out exactly what you are looking for to work with your freelancer or website designer to come up with exactly what you are looking for.

Make sure that you get a quote at the beginning of your project to know exactly what sort of price you are expecting to pay. This will help both people to be on the same page, and if something looks like it's costing more or less, you will know what's happening.

Websites can be expensive money pits once you go down the road of building your own. Ask for references or examples of people's work, and have a look at the work they have done for yourself. If possible, speak to the website administration to learn for yourself how well the freelancer worked out.

Remember to keep in mind everything you have learned so far about branding and keep your brand flowing through the website.

Keeping Communication Lines Open

One of the most important aspects of any working relationship, whether on a freelancing site or in real life, is communication. Successful communication with the people you work with and the clients you work for is an important way to establish an excellent working relationship. Obviously, you don't want to harass people, but keeping people up to date and letting them know that you have received their message or email is critical to your business relationship.

When you are working online, you don't have opportunities to develop a relationship the same way people do when dealing with clients face-to-face. Working on your online communication skills will help you secure more work and help keep those clients coming back time and time again.

There are millions of freelancers online, so any advantage you can get that puts you ahead of the competition is a good idea.

Always Use Proper Language When Speaking With Potential Clients

When we communicate with our friends and family, it is easy to fall into the bad habit of using slang or abbreviations. It's never going to be appropriate in a business relationship to use any abbreviations such as *Ty, gr8, 4u,* or yw, for example. After working with clients for a long period, you may use less formal means of talking but always try to maintain a strict business relationship. Speaking with freelancers about how they speak to potential clients, I have learned how just using correct grammar and spelling has won them jobs from among dozens of other freelancers, some more qualified.

Replying in a Prompt Manner

Some freelancing sites will evaluate and rate you on how quickly you reply to messages that you receive from prospective clients. It isn't normally every message, just the initial message from a new client. Fiverr, for example, has your response time displayed on your profile when potential clients browse your profile. If your average response time is less than a few hours, it could lead to new clients. Many people would much rather engage with someone they expect to answer within several hours instead of several days or a week.

You have to be realistic. You are not always going to be able to answer every new message within seconds, but even if you're busy, it isn't a lot of hard work to simply reply and say thank you for your inquiry, and I'll get back to you as soon as possible.

If you feel like you need more time to give them an accurate quote or answer a question, then let them know that. You don't want to rush into committing yourself to something that you either can't deliver or that will end up costing you time and money.

Don't Let Any Communication Get You Frustrated or Flustered

Sometimes, you are trying to deal with someone who may have a different first language than you or are just plain hard to understand. Just because you think about your responses carefully doesn't always mean that potential clients think about what they want or how they express that. Try not to get frustrated with them, as you never know who will be an extremely beneficial client.

Be clear on how you explain your point of view to them. Don't try and overcomplicate things if they have a problem with language. If you think it's worth it, try using a video or a screenshot to emphasize your point so that they understand what's going on.

Sometimes, it may be better to explain to the potential client that you don't have time to undertake their project or that you aren't suited to it. Quite often, if you let people know that you won't be able to do the best job possible, they are happy to move on. Sometimes, this is easier than dealing with a miscommunication between you and your client, which could end up with you having to either redo work or get negative feedback.

Always Get Written Confirmation of Any Changes

Suppose you speak over the phone or Skype with your clients or any other verbal communication. In that case, it always pays to get everything in writing. This doesn't just apply to changes either; all commitments should be recognized with a written agreement. If you take notes during the call, you can simply write up a brief report and send it to your client to get confirmation. This is a quick and easy way to get everything out in the open, with no opportunity for any misunderstanding later on.

If You Don't Understand Something, Then Ask

If you have ever dealt with a freelancer or employee who nods and says yes to everything you ask, and then halfway through the project is completely lost or delivers something entirely wrong, you'll understand how frustrating it is. There is nothing worse than to get not only something below average but also something that is not even close to what you expected.

You don't want to be that freelancer; your negative feedback will soon lead to your rating going down the toilet.

The only way to find out more information is to ask! The best time to ask is before you start a project. This will help you avoid any unpleasant consequences later on in the project. No one will be annoyed if you ask a lot of questions at the start of a project, especially if you deliver a great product in the end.

By following these simple steps for successful communication, you will be that much closer to having a great freelancing career!

Setting The Right Rate

The nice thing about skilled freelancing work? It pays *really* well. Why? Because, for most people and companies, hiring an independent contractor is cheaper and doesn't come with the kind of overhead that an employee comes with. That flexibility is worth it to them, and because they want the job done right and on time, they'll pay even more for the right freelancer.

Skilled freelance jobs that require creativity pay even better because **your creativity is valuable.** I think because our society devalues different forms of art (writing, design, illustration, etc.) as career paths, so many people are discouraged from pursuing these paths. Yet, we need artists and those with creative skills—marketable, regardless of what others may have told you. It might be less glamorous to write copy for a marketing campaign or create an animated video for a commercial, but these jobs can be fun in their own right and often pay pretty well.

So, part of being paid what you're worth is to know that you've got to set your rates to reflect how valuable your creative contributions are. Then, there's the financial reality of needing to know the bare minimum of what you'll need to get paid to pay the bills and have time to pursue your art.

I get a lot of potential clients to reach out to me on my blog asking for me to work below my rate or to forgo my rate for credit/equity on the

book or finished project. It speaks to the fact that most people misunderstand why freelancers charge what they do, and freelancers won't work below our rates.

I understand people wanting to try and negotiate, but clients need to know that there are freelancers that charge a wide range of rates based on their experiences, and you can search through freelance platforms to find the freelancer who fits your budget.

So let's shed some light on what goes into factoring in a freelancer's rate.

What's in a Rate

When you're freelancing, there's so much that you need to think about when formulating how much you need to earn. Here are a few of those things:

Education and Skills Training. Freelancers have paid or are paying for past and current education and skills training. Whether that's in the form of student loans, continuing education classes, or other kinds of workshops, books, or courses, learning never ends. You have to factor this kind of training into your rate to continue to deliver quality deliverables and continue to hone your craft.

Healthcare, Vacation, and Sick Time Off. Despite being remote workers, you can't *always* be working. Your rate has to be high enough so that you can take time off when you're sick or burned out. You also have to pay for health insurance out-of-pocket -and this is often much higher since our deductibles tend to be sky-high, and things like dental and vision can add up quickly on top of our health insurance fees.

Unpaid Administrative Work. Just like any business, you have to deal with the administrative side of the work. This means unpaid calls and emails with potential new clients, submitting proposals for new jobs, updating your website and portfolio, submitting invoices, etc. Unlike traditional full-time employees, you don't get an annual salary, so this work goes unpaid, and you have to factor this time into your rate since billable hours often don't cover it. Sometimes, you may need to bring on a Virtual Assistant to help with the heavy lifting of invoicing, scheduling, and other admin tasks, but that's more something to consider when you're a full-time freelancer juggling multiple clients and projects at once.

Taxes and Fees. You have to factor freelance taxes into your earnings, as well as the fact that platforms like Upwork take 10-20% depending on how much you bill with a particular client. You often have to hire CPAs since freelance earnings can be complicated, and tax deductions can be tricky to navigate.

Equipment, Tech, Phone Bills. Depending on your work, you need equipment and things to do your work properly. Whether that's something as simple as a laptop, smartphone, and super-fast internet, or more complex like camera equipment, these are all business expenses that you have to factor into our rates. If you're making international calls with clients, you need to factor that into your phone bill as well.

Marketing Costs. This could cover something like website hosting, or if you need to market your services, things like Instagram or Facebook ads. It also includes things like business cards to hand out at networking events or mixers.

Your "Office" and Additional Utilities. If you have enough space to work from home, that's great! You can avoid renting out an office space. However, you still have to pay for the additional utilities from being home all day that you wouldn't be paying if you were working a traditional office job instead.

Living in Expensive Cities. Because I work in TV, I have to live in Los Angeles. This is also sometimes why my clients hire me: I'm plugged into a unique, artistic community and am working in Hollywood. They value that experience -- but it comes with a price tag. LA is not cheap, and the cost of living needs to be factored into my rate, and you should factor the cost of living into yours, too.

Keep in mind energy management **--** not just time management -when you're setting rates and organizing your schedule. You probably won't be putting in 40 billable hours a week every week (depending on your client base and type of freelance work.) So much of that time is taken up by administrative work, marketing your services, emailing clients, sending invoices, etc., that 40-80 hour workweeks aren't necessarily the norm depending on the field you're in.

If this is intimidating — don't panic! You can start at a lower hourly rate, and then once you gain experience, clients, and testimonials, slowly raise your rates.

That's why **it's** better to start off freelancing part-time before making the leap.

Part-Time Versus Full-Time

I want to take a second for a bit of a reality check here: it took me about two years of freelancing part-time on top of a full-time job to build up a freelance portfolio and client base that allowed me to take the leap and go full-time. And by that point, I started making money through my creative endeavors, which allowed both my freelance and creative income to dovetail and pay the bills so that I no longer needed to hold down a day job or assistant job.

It's not a science, and you'll want to do your best to save up and build your client base before you dive into a full-time freelancing and creating lifestyle, sans day job.

Everyone's finances are different, so it's up to you to take a realistic inventory of your expenses and where you're at before making the leap. It's much harder to go back to a day job once you've left, so if you're in a good spot, try and hold out for as long as you can before you quit.

You probably have heard about diversifying your portfolio when it comes to investments, but you should diversify your freelance income as well. We'll get into that more later, but for now, let's talk about the different ways you can structure your own rate as a freelancer.

Different Types of Freelance Rates

The nice thing about the freelance platform Upwork is that you can snoop around other freelancer's profiles and see who's charging what and what their experience looks like.

Before you set your rates, **see what others are charging,** and then **use a freelance rate calculator** to calculate your expenses and what you need to make to pay the bills. I hate math so go to Google to go find one of those.

You'll also need to consider **fixed-rate** versus **hourly:**

You should charge a **fixed rate** for projects with super-clear deliverables and guidelines. You need to be upfront with your client when it comes to the timeline and scope of this project, and

I recommend charging half upfront and half upon completion. If your project requires rewrites or revisions of some kind, consider creating milestones for every revision (i.e., after every revision, your client pays you a portion of the fixed rate, and there are a fixed number of revisions per project, and additional revisions are extra.)

Use an **hourly rate** when you're not entirely sure how long a project will take or if there's a lot of back-and-forth revisions and complex aspects to it. This is great if you have clients who want to work at their own pace with you and need a flexible option when it comes to their project.

If you're doing a long-term project, consider negotiating a **monthly fixed rate** for your clients to retain your creative services. For x amount of money, you'll be working on their project and setting flexible milestones on progress as you go, working an estimated number of hours of them per month. This is great for long-term projects because you can allocate the bandwidth you need to a project and have an expectation that you're going to be doing a certain amount of work for that client per month. This can be better than hourly fees because hourly clients can often disappear. It's hard sometimes to get your work schedule on track to hit your income goals if you don't have some sort of steady monthly income. However, it's a less flexible option for your client, and some prefer hourly instead.

How to Get Paid

When I was just getting started with freelancing, I sheepishly Googled: *"What is an invoice?"* It's okay not to know the answers to these things! When it comes to the freelance basics, an invoice is what you send to your client that shows how much they owe you for the work you've done and how to pay you (via Venmo? Paypal? Check? If so, who should the check be made out to and where should it be mailed?) You can find invoice templates online or in Apple Pages or Microsoft Word. You'll need to address an invoice with an *Attn:* to the person who is going to pay you at whatever company they work at if they work at a company.

How often you invoice a client is up to you and your payment schedule. For me, I send invoices every other week encapsulating the number of hours I worked during those two weeks and generally what I was doing for those hours. For example, the line items on my invoice will read things like: "Call - 4 hours" and then the amount (4 hours x my hourly rate, with all the line items totaled up at the bottom of the invoice.) You should also list when you expect to be paid on your invoice, such as within 30 days upon sending the invoice. You can send invoices weekly, bi-weekly, monthly, or only after certain milestones are completed in a project. For hourly projects, I'd recommend sending them every other week or even weekly just so your clients don't get hit with a huge bill every month. Breaking things up in a more piecemeal way feels more manageable, but this will vary depending on the client and project.

You also may need to keep a **work diary.** Some of my clients like to see a breakdown at the end of the month of how I spent my working hours, so I'll create documents that specify the tasks I did for them, any meetings we had, and how much time each of those tasks and meetings took. Different clients will want different levels of specificity in your work diary, and some won't need you to keep one at all. Even if my

the client doesn't ask me for a work diary; I still break down my hours in a more general sense on my invoice and can answer any questions that might come up in terms of how I spend my time.

If a client refuses to pay for your work, you should stop working immediately. From that point on, you should continue to follow up and seek payment without rendering any additional services. If a client is struggling or refuses to pay you, that's a huge red flag. You don't want to waste your time and energy. You need to take your work seriously as a professional, so I would recommend stopping working for that client immediately. Depending on how much you're owed, you can seek legal action, but it won't be worth it most of the time. Check with your CPA, but you also may be able to write off unpaid invoices on your taxes.

I like working on Upwork because invoicing, work diaries, contracts with clients, and getting paid are all automated through their system, and you get paid out weekly. However, I have a lot of clients who prefer not to work on Upwork, and because Upwork takes a percentage of your earnings, I'm okay with working with clients not on Upwork as well. However, suppose you find yourself getting burned by clients not paying you or feel overwhelmed with all these processes. In that case, you may want to consider working solely on a platform like Upwork. More on that later in this book!

At the end of the day, you'll need to work with your clients to figure out a pay structure everybody's cool with. The way you approach getting paid for your work will evolve as you learn, so for now, set your rates, pick some preferably legal activities you can charge money for, and get yourself out there in the world.

Managing Your Client And Projects The Right Way

I write this chapter with love: I truly do respect all my clients. I wouldn't be working with them if that wasn't the case, and I have dropped clients who were disrespectful or weren't worth working with or took up my time and energy in a way that just wasn't the right fit. You'll know those clients when you run into them. This chapter isn't about them, though; it's about **juggling client expectations** with the **complex workload** that comes with being a freelancer—and staying true to your identity as a creative, with your own pursuits on top of freelancing.

When I was a part-time freelancer while I was also an assistant, some potential clients wouldn't want to work with me since my schedule wasn't wide-open. I had to take calls on my long drive home at 8 pm (yay, Hollywood hours) or on weekends or early mornings. Yet, when I was part-timing as a freelancer, I had the stability of a full-time job to fall back on. When you go full-time for your freelancing and your artistic work, you lose that stability, and you have to do a whole lot of math when it comes to how much you need to make so you can take a sick day or a vacation or cover the cost of your own health insurance.

Your clients don't see that side of it, of course: when you're a full-time freelancer, they assume you're just doin' your 9-to-5 hustle and hanging out at your desk waiting for them to call. Some clients will know you're managing other clients but still probably don't under

stand the kind of bandwidth juggling multiple clients take. Still, others will expect you to work weekends or weeknights. When you get a call at 9 pm on a Saturday from a client who wants to chat about your project —well, that's not okay. (Unless you're okay with working that late on the weekend.) This is where communicating deadlines come in.

Before all that, though, you need to iron out your project tracking board and get organized, so you don't miss deadlines or get overwhelmed. Then, you can start creating boundaries and communicating those with clients while still being sufficiently available to them so you can complete your work and collaborate with your clients effectively.

For the Love of God, Track Your Projects

At any one point in time, I could be in the middle of 7 or **17** projects. Honestly? I don't know sometimes. I have clients that pop back into my life after a year of... I don't know, sailing the seven seas? Starting a new business? They lead busy lives, man. My freelance projects (especially the smaller ones) can start and stop sporadically, and I need to pick a project back up at any time.

What I'm trying to say inarticulately is that your clients have lives, too. They'll sometimes peace out for weeks... months... sometimes years. Then, they'll resurface and be like, HELLO, LET'S CONTINUE TO DO THE THING! and you'll be like yes! Of course! And then you'll need to figure out how to shuffle them into the mix of your current clients because one of your current clients could also decide to go on a hiatus. You're left trying to figure out how to pay bills. (This is why income diversification is key. More on that later, friends.)

Anyways, you need a way to track your projects and clients. Organize your goddamn computer files. Keep all your notes where you can easily find them. Store emails in handy folders organized by clients. Don't just scribble notes and lose them to the abyss of your desk drawers!

I recommend using a combination of the following free apps and websites:

Trello is great for tracking individual projects in a board-like format. You can get really colorful and pick a background. It's a good digital equivalent of notecards on a bulletin board. It has some great team collaboration features as well!

Wunderlist is my favorite to-do list out there, and you can set individual deadlines and make shared lists with others, also! They have a great mobile app that is the best one I've found in the to-do list space.

Google Drive is a key component of any organized person. Still, if you don't use this by now, I'm just going to assume you've been living under a rock in Area 51. (In which case, how are the aliens? Are they cool?) I like putting memoir manuscripts in a Google Doc because my client and I can collaboratively leave notes on the draft and go back-and-forth in discussion over it. Too many people use Microsoft Word (ick), and tracking revisions on that SUCKS. I much prefer Google Docs, so we don't have to worry about formatting turning to sludge when sending documents back and forth and exporting them from Apple Pages > Microsoft Word.

Evernote is what I use to store all my notes from every call *ever* —from current, prospective, and past clients. Bigger projects get their own folder, but small projects get individual notes within my freelance project notebook.

Dropbox is great for sharing deliverables effortlessly, and I sometimes create shared folders in there if I'm going back and forth with a client on certain things (but don't work collaboratively on a document if you can help it, Google Docs is superior for document collaboration.)

There's a zillion more if you feel like color-coding the hell out of your world, but just please pick a system because life is stressful enough without having to remember which client told you which anecdote about their childhood.

Deadlines Are Not Your Friend

This is going to sound weird, but... I try not to set deadlines. I'll give my clients estimates on when things will be done, but I try really hard not to set hard deadlines. (I'm not always good at this. It's still a work in progress.) In my head, I'll know what milestones I need to meet when. Still, I try not to communicate those specific days to my clients unless it's an intensely collaborative project with many moving parts. I have to set deadlines so other people can know when to expect things.

Why do I hate deadlines? It's not because I'm lazy. It's because of the shuffling above. My schedule is fluid and constantly shifting to accommodate all my different clients, life things, and other businesses. My clients hire me because I'm doing cool things—and it is sometimes those cool things that pull my attention away from my freelance work.

So, I try not to set deadlines, which helps my stress levels. That way, I don't feel like I'm failing, which keeps me in a positive mood, and ironically helps me finish things faster than if I did have deadlines.

I had occasionally set aggressive deadlines when I thought I needed the motivation to get a lot of work done quickly, but it often backfired leaving me burnt out or exhausted. In my opinion, it's better to work intuitively while keeping in mind your client's broader timeline and then communicating with them along the way. You'll need to figure out your own relationship to deadlines, but basically, if you have to set one, give yourself more time than you think you'll need. That way, you can deliver your work early in the best case and on time in the worst-case.

You may also need to remind certain clients that you've got other projects. If they wanted your exclusive attention, they'd hire you as an employee with a full-time salary. Assuming neither of you wants or can have that relationship, they'll need to understand why you have to stay flexible. There are even laws about this in certain states regarding the classification of freelancer versus employee: an employee is required to be somewhere at a certain time, and a freelancer sets their own schedule (or some verbiage similar to this.) As a freelancer, your clients need to know you've got to keep up a bunch of other projects and clients to pay the bills.

Short Term vs. Long Term Projects and Clients

When you're building your client base, it's important to try and find the right balance between **short-term** and **long-term** clients.

Long-term clients are great because you can build fruitful working relationships with awesome people, and you'll be able to get into a rhythm of knowing how to work well together. I really love some of my long-term clients, and even when I jump into writers' rooms, I will still try and carve out some time to keep up with these clients if it makes sense for my schedule and their deadlines. On the other hand, if you make a long-term commitment you can't or don't want to keep, it can be uncomfortable to disentangle yourself after your clients have already gone down the road with you. There's less flexibility with longer-term clients. If you have to pass up other opportunities because of the amount of bandwidth the longer term clients take, you may find yourself in a tough spot if your long term client ends their working relationship with you. You need to scramble to fill in the gaps. However, long-term clients are a key part of your freelance business because the amount of time it takes to find and onboard a new client can be inefficient. This is one reason I got into memoir writing: I work for clients for months at a time, which means I know I'll have steady work.

Short-term projects like blog posts are much more start-and-stop, and it's harder to predict your workload and plan and budget accordingly. Shorter projects also afford a kind of flexibility that long-term projects don't have: you're able to hop on and off projects more easily, and it's no big deal if you end up not having availability or interest on a short-term project. It's more of a hassle to find and vet new clients, so you'll need to charge a bit more with a short-term project because there's a good amount of unpaid time you'll need to spend finding and interviewing for these projects.

As a freelancer, you'll want a mix of both short-term and long-term projects to **reduce risk** and be able to pay your bills without giving up your independence. If you only have two long-term clients, and then all of a sudden, one stops working with you, you have to quickly fill that void with a new client to pay the bills. (Ideally, you'd have saved enough money to tide you over.) The more clients you work with, the more your risk is spread out across all of your clients: chances are, they're not all going to leave you at the same time.

As a creative, it's also key to **retain your flexibility** and have different kinds of projects and clients. Suppose you are commissioned to paint a painting or hire into a writers' room or direct a film. In that case, you have to shuffle things around to make time for new opportunities to grow as a creative person. You want to position yourself so you can go after all the exciting creative projects that come your way: freelancing should be seen as a rewarding way to clear a path toward creative freedom. When to Stop Working with a Client

Occasionally, you'll end up working with a client who ends up not being a great fit. Sometimes, your work styles don't mesh, or they don't share the same values as you—or you two just can't get on the same page. In this case, you'll need to find a way to gracefully bow out of your working relationship with them and start fresh.

As a freelancer, your reputation is important, especially if you're working on platforms like Upwork, where your client reviews are posted publicly. So if you've made a commitment to a client, try and finish out the project. It's best to follow through and then end your working relationship when your commitment naturally comes to an end.

However, sometimes a project is dead on arrival, and it's super clear that things aren't going to work with you and your client. In some cases, I've refunded a client and parted ways amicably, which was best for both of us: they still left me a nice review, and I was able to preserve my energy and refocus on the clients I was really excited about.

Some clients may try to take advantage of you and demand refunds that you can't afford to make: in this case, you need to stand strong in the work you've already done. This is in part why I mostly work hourly: my time is money, and when a client hires me, I deliver work that I need to be paid for. An hour you work for one client is an hour you could have spent working for someone else or working on your creative projects, and you have bills to pay.

During a project, check in with your clients periodically. Ask them how things are going and if they're satisfied with the work done so far. If they're not, then offer to bow out and recommend another freelancer instead of just powering through. If you can, offer partial refunds; if you can't, maybe offer to do a little bit of additional free work.

There are rare cases in which you might find yourself working with

a toxic client who doesn't respect you. In those cases, drop them and run like the wind. Try to do so with as much grace and kindness as possible, but also don't allow yourself to get entangled with people who don't treat you well.

Freelancing should be a positive and rewarding experience: don't work with people who don't respect you and your work.

Setting Boundaries and Enjoying Your Free Time

The nice thing about working freelance? I can go on an 11 am hike in Los Leones Canyon on a Tuesday and won't be mobbed. I can shoot some hoops at the park on a Thursday afternoon with another friend who's working from home or go see a movie if I feel like it. The downside? I'm often working on weekends or odd hours.

I've gotten a lot better at this. Now, I'll only spend an hour or two on Saturday and Sunday morning responding to emails and doing some freelance work. I used to blow through my weekend trying to meet deadlines, taking calls, and treating weekends like weekdays. That was a mistake, and I've finally started creating the boundaries again.

As a creative, you need your weekdays to explore, hone your craft, or take in new experiences. The whole point of the freelance lifestyle is to give you room to breathe and find inspiration in unlikely places. If you're suffocating yourself at work and not taking full days off, you're on the road to burnout. You're also not allowing yourself and your creativity room to breathe: you might as well have a normal day job in that case!

If your clients need you to be available on weekends, try and sequester work time to something like Saturday morning, leaving the rest of your weekends free. Otherwise, see if you can take Friday and Saturday off and work Sundays. No matter what configuration you end up with, know that you actually do need rest—so take it! Turn off your phone, tell your clients you're unreachable, and get off the grid.

Basically, client and project management comes down to designing your workdays and workweeks to be productive and get the rest and necessary downtime so you can still connect with your creativity and your own projects. But how can you make sure you're finding that balance?

Managing Your Energy And Time

You're in the swing of things now: you've got a full client roster, you're building new contacts, and you are working on several projects at once and paying the bills.

So how do you manage your time and energy so that you can freelance and make progress on your creative projects?

Building Your Daily Schedule

I'll use my schedule as an example, but keep in mind you'll need to experiment to find what works for you.

Full-Time Freelancing Schedule:

8 am. ***Clearing space.*** I wake up, make coffee, clean my apartment, and prep my workspace. I have a daily planner in which I'll write down my priorities for the day and figure out if today is a Freelance Day or Creative Day. Since this schedule is for my Freelance Day, I'll look at the deadlines coming up on my calendar, add any relevant tasks to my day, and try and create big chunks of time for each project. If I can, I'll try to devote an entire day or half-day per week to a particular client, so I'm not switching constantly between projects. Before I dive into my day, I'll answer emails before closing out of my email and any other distractions. This first hour of the day helps me plan and prepare for my day while closing the loop with any emails or text messages I may not have gotten yesterday.

9am-12pm. ***Work session #1.*** From 9 am-12 am, I'm in serious work mode. I'm focused on my top priorities and working quickly. I know I have the most energy early in the day, so this work session is my most productive.

12 pm -up. *Break for lunch and work out.* In the afternoon, I'll take a break to eat lunch on my balcony and work out. Generally, I try to stick to 4 days of HIIT weight and cardio training per week, mixing in hiking and yoga 2-3 days per week. Workouts are important: if you work from home, it can be easy to get lazy and not move. Use the time you would have spent commuting to and from an office and invest it into your physical health. It will make you feel more energized and be more productive!

1pm-3pm. *Work session #2.* My second work session is less productive as I slide in the afternoon, but here I try to wrap up the priority tasks I need to get done and also jam through emails and communications I need to answer. I have about 6-7 hours of great work in me per day, and after that point, the returns start to decline.

3pm-6pm. *Break OR meetings and phone calls.* In the afternoon, I try to take a break, go for a walk, read a book, get outside, or take meetings and phone calls. I try and never schedule client calls in the morning because that takes away from my productive work sessions. In the afternoon, I'm no longer planning on getting real work done, which makes it primetime for meetings, socializing, and answering emails, or doing lower priority tasks. I'll also try and squeeze in an early dinner.

6pm-10pm. ***Drinks, networking events, classes, shows.*** It's no secret: I'm a night owl. In the evenings, I'm out socializing and networking or at concerts or immersive theatre shows. Sometimes I take dance classes or go to book clubs or writers' groups. Basically, I have almost all my evenings blocked out for fun things, which is intentional. I try not to interrupt my days, stacking all of my work in a fairly traditional workday so I can enjoy my evenings. If I'm on a deadline, this time slot will turn into a third work session, but I really try to avoid it because I get burned out if I put in too many hours each day. I work fast and get a lot done, which means I can get more done in less time, but also, my energy levels get used up pretty quickly as well.

Amy's Daily Schedule for Creating:

10amish. ***Wake up.*** Sleep is part of my creative process, so on my creative days, I sleep in for as long as I want. Moving slowly is key to allowing thoughts and ideas to percolate instead of forcing them to happen.

10 am -Whenever. ***Writing session.*** I'll usually write for 4-5 hours at a time and then take breaks to walk, do something fun, or watch or read something to give me inspiration. There is no set schedule. It just kinda happens. When I'm rewriting a project, I'll print out the script or manuscript I'm working on and spend the day with colored index cards and various art supplies, covering my room in notecards and color. I'll lie on my floor and listen to my writing playlists to get in the mood. I have writers' groups and other methods to hold myself accountable, so I'm not just living in a daydream, but the key here is to ***make the space your art demands of you.***

If I don't have the luxury of taking full days off for your art, try and protect pockets of time—no matter how small—to work creatively. I spent every November for six years writing a 50,000 novel for National Novel Writing Month, and that taught me that creative work could be created anytime, anywhere. Obviously, it's an ideal situation to be able to abscond to a cabin in the woods where your Mom is doing your laundry. At the same time, you wax poetic about society beside Walden Pond. Still, not all of us have access to cabins or parents to do our laundry. Do what you can with the time you have.

Energy Management

When you're building your daily schedule, it's critical to keep in mind energy management.

When you're working a traditional 9-to-5 job, it's okay for you not to have productive hours. Chatting with co-workers, taking breaks, and surfing the internet is part of working hours usually, and you still get paid for that time.

But when you're a freelancer, you don't get that time to goof off or take a break on the company's dime. That's in part why we charge higher rates. We're often cheaper for companies because we don't come with overhead costs associated with traditional employees, but we have to make up for some of the benefits lost in our hourly rate.

It can also be unsustainable for us to bill a traditional 8 or 12-hour workday, depending on the kind of freelance work we do.

Not only does this all have to be factored into our rates, but it also has to be factored in how we organize our day so that we don't get burned out and can still have energy left over for our creative pursuits.

Here are the different aspects of freelancing that sap our energy:

- Difficult clients
- Complex projects without clear guidelines or expectations

- Poorly managed projects
- Juggling a large number of clients
- Work that we're not passionate about
- Fast turnarounds or unrealistic deadlines
- Low-paying assignments or clients

Some of these factors are out of our control, but some are controllable. Suppose a client ends up being tough to work with and affecting your mood and productivity to the point where you're falling behind on your other work. In that case, it might be time to let them go—even if they pay well.

Same with the opposite: if you work for a really great client but don't pay you well, you might need to move on if you're having trouble managing your workload and your other clients pay you better.

The **type** of work you're doing matters, too. Once, I had a client who paid super well, and I could always expect an oversized paycheck from them every month. It was nice until I found myself dreading the work I was doing. I found myself struggling to meet deadlines and feeling exhausted every time I worked on that project. I ended up parting ways with that client—and the space I made was suddenly filled with work I was actually excited about and actually ended up paying better, too!

While it can be scary to turn down work or walk away from a lucrative assignment, pretty much every problem you encounter as a freelancer can be traced back to **time** or **energy management.** If you feel burned out consistently to the point where a day off isn't helping, then take a critical look at what makes you feel drained, and make adjustments to your projects from there.

Staying On Track Creatively

What are your goals as a creative? What are the projects that matter most to you? What do you want to accomplish this week, month, and year?

The creative process often happens on its own schedule, and with the demands of freelance work, it can be easy to put your creative work on the backburner. Try not to let this happen. If you can, join writers' groups or accountability groups so that you can stay connected with your creative work. Challenge yourself to do *one creative thing* every day, or at least every week. Write a page every day. Do a sketch every

week. Whatever your art form, try and keep connected to it even as you're finding your footing as a freelancer. Be gentle with yourself, but also never lose sight of what you're most passionate about.

Separating Creative Work from Freelance Work

This is a really important tip nobody tells you.

Your freelance work needs to be separate from your creative work. Read that again. And again.

As freelancers, we may use creative skills to do our freelance work. However, our pure creative endeavors need to be divorced from our breadwinning freelance work.

When I'm writing a script or novel, I need to slip into my characters. I need to meditate on what truths I'm trying to capture on the page or on-screen. I need to approach the work with reverence and respect and honor what I'm putting out in the world. This process takes time, dedication and **cannot** be done solely for money. A part of your soul has to come alive when you're creating. It's a magical flow state that has to be accessed from a special headspace.

When I'm doing freelance work, it's much more formal and structured. I have calls, deadlines, emails to respond to, clients to cater to, and I also need to juggle different clients and projects to pay the bills. I have to negotiate, invoice, follow up, and handle incoming leads. It's a business, it's work, and while I really love my freelance work, my job is to deliver something that my clients love at the end of the day.

When I'm freelancing, I'm writing for someone else. When I'm writing creatively, I'm writing for myself.

Sure, I'm also hired by other people for my creative writing, but that's a different kind of work. I think that professional creatives need to be able to retain some sort of autonomy and voice and also create boundaries around their work, so they're still doing it for the sake of their art. Becoming a "sell-out" is bad for your soul, which is why freelancing helps you retain the financial freedom only to take the creative projects that catch alight for you.

When you're designing your schedule, try and **separate your creative time from your freelance time.** This will ensure that you don't see your creative work as "work," and you're able to carve out the space you need to breathe creatively.

Freelance work often moves at a fast pace, whereas creative work sometimes needs a slower pace. Take a look at the pacing of your week: I try and stack my freelance work at the beginning of my week, where I'm able to move fast and knock out a ton of things at once.

This leaves the second half of my week fairly free to devote largely

periods of reading, hikes, doing some freewriting, and working on creative projects. My willpower is winding down as the weekend approaches anyways, so doing fun things helps me get back in touch with things that inspire me. Then on the weekends, I'll set aside my freelance work and just focus on hanging out with friends, having cool experiences, and writing stories.

Take a look at your workweeks: how can you experiment with how you stack projects? How can you carve out time to be creative? If you find yourself starting to get overwhelmed with everything you need to do, hiring a virtual assistant or outsourcing some of your tasks may make sense.

Chapter 7

Making Online Course To Make Money

The Basics Of Teaching Online

At the moment, I'm writing; I have four online courses in the pipeline. They are online video courses that I intend to deliver shortly to my audience, complementing the other 17-18 courses already created.

Someone occasionally asks me how I can create so many online courses for my clients, how long it takes me, and how I can continue with my training at the same time. It's not like I'm a super-fast typist or a guy who shoots 10 hours of video and then gets the footage edited by some collaborator; absolutely not.

The same goes for my training; I organize it according to precise and scheduled appointments: when you study a course online, you just do it as if it were an appointment with a client.

In this way, I can efficiently manage every commitment, not to mention that I produce is thought, created, and finalized only and exclusively by myself.

Another important consideration, which I came to some time ago, is that today it does not always make much sense to create giant online courses with the risk that no one will ever digest them.

Suppose I look back at some of the programs I created 5 years ago. In that case, I see with regret that only a small percentage of paying students have completed the longer online courses.

The Problem of Online Courses

I live on paid content, it's 50% of my business, but it doesn't have to take 50% of my time, and it doesn't have to be the same for my students. Time is the most scarce and non-renewable resource we have. From Mark Zuckerberg to the postman who brings you Amazon packages, everyone has the same time and the same impossibility to renew it. Every day there are 1000 things to do. This has briefly created the misperception that many entrepreneurs have of their work, concluding that they have no time to train.

More and more, I receive requests from professionals and entrepreneurs who think that time should only be spent working with their heads down, without understanding marketing, with the result that everyone wants to delegate, only to realize that they do not have the money to do it and that they do not know what is being done for them.

With less and less time available to study and more and more the need of not staying behind, I have long since realized that each online course can be structured in two ways:

Super-fast online courses to consume in 2-3 hours.

Super structured online courses, perhaps with 30 hours of video, but extremely modular, which you can access if it is necessary to consult the part that interests you.

In the first case, these are extremely specific paths, which go to hit and solve a single but important problem, with a quick solution and without having to come to tell the life and times of each topic.

These are also the easiest courses to sell because the objectives, the results, the transformation, the oppositions, and the target audience are clear.

On the other hand, we find the online courses more structured and richer in content; they are more complex to implement and perhaps a little more difficult to sell but usually have a much longer life. In this case, the audience's request and the promises are very different from the first type seen earlier. This kind of online course includes strategies that last over time and offer a multi-disciplinary approach.

Here the topic is more impressive than the ultra-fast courses. It starts from a more general assumption to arrive at a well-defined result, such as monetizing socials.

This kind, of course, is approached by people who have very serious intentions and less urgency; they are entrepreneurs who want to train and build a solid and lasting foundation, which they will draw from there for years to come.

They want to create a permanent culture on the subject and maybe develop their own point of view.

How to Make an Online Course

There are several ways I have used in the past, and, for one reason or another, they are all extremely effective.

Let's see them together quickly. I'll go more in-depth later on.

Video Course in Room. A classic, go in front of the camera, and talk to your audience, perhaps with the help of a digital blackboard or flipchart. Explain your content by looking at the lens and talking to your student as if it were right in front of you.

Here you need a couple of good lights, a microphone and a video camera, a reflex camera or a smartphone. All in all, things you can find quickly or that you already have.

Video Course with Slide. If you are afraid to stand in front of the camera or if your course requires screen display — tutor is for a program, for example— this is the best solution. I use Screenflow to do a screen capture and record all the audio.

Webinar. Why not talk in the camera or with slides live? That way, you can also answer your students' questions. Not only that, but this technique also allows you —if you structure the contents well— to divide the recording of your webinar into videos that will compose an online course.

I use LiveWebinar for webinars, but almost every platform you can find on the market today is fine.

Text Only. We go back in time a bit, before the advent of YouTube and streaming, and I really like that. Maybe it's because I like writing, but even more, reading and studying on content that I can quickly return to as many times as I want.

You can create a PDF or a sequence of emails or even have a course area where texts take the place of videos.

Audio. I'm a big fan of podcasts and audiobooks. This way, I can easily digest a lot of content while doing other things, like driving, relaxing with my eyes closed, magic...

If you're interested in Podcasting, I recommend reading the book *"Podcast"* by George Sims. It's a great guide for beginners.

Making an audio course is super easy. All you need is a good one. Microphone (the best low cost is the Samson Meteor on Amazon) and a computer, or your smartphone with some app to record voice notes and a good headset. The most important thing is the tranquillity around you.

You need to be in a quiet place to record, and if you're not super used to arm-talking, you need to prepare your texts for reading first. Not very suitable for tutorials and super practical courses. These modalities are all effective, and each of us will have our own positive and negative aspects.

Often the choice of the mode to be adopted for the creation of our online courses falls on what is more "in our style" or the technical skills we already have or the resources we can use.

In any case, nobody forbids us to make a "mix & match" of all these ways to create our online courses, for example by shooting a video course in the room, with some video slide lessons, an in-depth webinar, an appendix written and then recorded in audio.

When Is the Right Time to Start Creating Your Course?

In general, the beginning of the year is the ideal time for good intentions and also to define what you want to do with your first online course.

As for me, every moment is a good time to start.

There is no reason in the world to wait for good astral conjunctions or the perfect weather. Postponing doesn't help, and it's a defeat, even before you leave. So, if you don't want to regret it, you should start as soon as possible.

Setting a target, however, is not enough. In fact, *only 8% of people reach them every year,* and those are the people who have good habits and have worked hard.

Today I want to help you join the ranks of the successful ones. Are you ready?

Let's start with these eleven super practical tips to ensure the success of your goal and finally see your online course launched within this year.

Start With Concrete Goals. It's impossible to achieve a goal if you don't know exactly what success will be like. If your goal sounds like "Make a lot of money in 2020", what will meet your definition of "a lot"?

Instead, think of a concrete number to target. Whether it's 2,000$ or 200,000$, write it down and keep that number in your head.

Note: It doesn't matter what level you're at in your business because if you've never created an online course, the difficulties you'll encounter and have to overcome are probably very similar. David is Goliath, so don't be intimidated, okay?

Also, be sure to add an interval to your target. There is a big difference between earning 200,000$ in a year or 10 years.

So, in summary, set a "reasonable but not too much" target because if you aim too low, you risk never getting off the ground and not exploiting 100% of the potential you have inside you, just waiting to be unleashed.

And this is only possible if you decide to dare and get out of your ADL (area of laziness).

Create a Precise Action Plan. It's easy to say, "I want to make an extra $10,000 this year" or "I'd like to start eating healthier." The reality —supported by a recent British study that I will perhaps tell you about another time— is that you are more likely to fail if you do not take any specific action that follows your goals.

When you set your goals to create a course, consider a simplified, standard formulation:

By a certain time, I'm going to: do X

do Y

do Z

This could translate into something like this:

In 2020, I was going to earn an extra $10,000 online.

Complete the creation of my online course

Launch to my list of 2,000 people

Hosting 4 webinars to promote my online course

Set Priorities to Achieve Your Goal. This is probably gonna sound pretty obvious to you. But if only 8% of people are achieving their goals, that leads me to believe that over 90% of people don't really care about their goals.

DO NOT set a goal because it's something you feel you have to do; instead, set goals where you can commit yourself 100% and put all your talents to use as if it were an ethical duty to achieve them and get the related benefits.

Cut out as much time as you can every day to pursue your goal, and you may be surprised at how quickly you achieve it.

If even 30 minutes a day may seem few, multiply them by 365, and you'll get over 180 hours total (and virtually free) to dedicate to the creation and sale of your online course: a real asset!

Make Public Your Goals for Your Online Course. It's easier to achieve your goals when someone comforts and supports you, whereas it's easier to abandon them if no one knows you've set them.

You can tell your mum, your partner, a friend... or you can tell them directly on the web, as having someone to report your progress and difficulties will help you to maintain the right focus and

motivation to complete your online course.

Find an Accountability Partner. If you want to improve on the previous suggestion, find someone trying to achieve a similar goal to yours (or has just done so) and work together to motivate each other and feel mutually responsible.

You can find this person anywhere from your community, social media, or private groups for online entrepreneurs.

If you need an extra level of motivation or want to make the process with your partner more fun, consider using an app like *Stickk.* The way *Stickk* works is to set your goal, add a financial penalty for not reaching the goal, and assign a referee to monitor your progress. If you don't reach your goal, the money you put in is donated to charity.

0 Car [2] [3] REFEREE [4] tksuppoRT

Put Processes in Place to Save Even More Time. Personally, before I became an expert on the subject, my biggest barrier to achieving the goals I set was always the mismanagement of my time.

When you have a million things on your plate, figuring out how to do everything can be daunting, and you end up doing nothing.

If you're focusing on creating an online course to build or expand your online business, find out what parts of your current business (if you already have one) you can automate.

For example, if you spend too much time on social media, find out what you can plan in advance. You can use programs like Edgar, HootSuite, or the great PostPickr to run your social channels on autopilot.

Defer to Someone Else What You Don't Have Time For. In the end, the most successful people have reached the point where they are focusing on the things they are good at and delegating what is not in their "area of genius."

You need to know your strengths when you have a business to run. If it's in your budget, look for a helper, and don't waste time doing what you don't like and/or are not super good at.

You can find virtual assistants in Facebook groups, search "VA =virtual assistant" on Google, and you'll find an avalanche of them) or you can turn to specialized sites like UpWork.com or Freelancer.com.

If you don't get bogged down in those parts of your business that are not the ones where you excel, you can focus your energy on true growth and expansion.

Celebrate Small Victories. If your goal is huge (and I hope it is), it can seem daunting when you're just starting out, and progress is very slow and not very visible.

But in my eyes, all progress is good progress. Even if you're only 1/10 of the way to your $100,000 target, think about all you can do with that extra $10,000 that you wouldn't otherwise have had if you hadn't started working towards your goal.

Remember this: it is better to reach a super-challenging $100,000 target at 50% than a "very realistic" $10,000 goal at 100%.

I'm telling you this from personal experience. It can be exhausting to manage your online business. Don't let small winnings go unnoticed. Celebrating "mini-milestones" along the way can help you maintain your motivation and focus on the ultimate goal without getting caught up in anxiety and frustration when the goal is still a long way off.

Here's how to start organizing a step-by-step online course, giving the right structure to your project.

Divide the creation of the course into steps, for example:

- Finish writing the content
- Complete the first module
- Finish a chapter

Then, you put in both a deadline and a small reward to get it done quickly and without procrastinating.

Re-Concentrate When You're Facing Some Difficulty or Emergency. More often than not, you'll face some setbacks on

your way to the success and completion of your course. Especially when it's your first attempt (or after you've already taken other tutorials or paid courses without success), it's easy to give up when difficulties arise, but often the problems you encounter can help you understand what's wrong.

1....Suppat Harkoprg AccauntIn. Sao Ali Cktkpa~ I'll tell you more: when you are faced with new problems, it is because your level of knowledge/experience has increased, and these problems are the concrete result that you have passed the level. So take them positively and worry only if things always go smoothly.

For example, if you launched your course or your book and completely failed, instead of throwing in the towel, reevaluate everything. Review your sales letter, your email marketing strategy, or your mailing list. Find out what went wrong and what changes you can make.

But there's more.

Finding yourself having to deal with an unexpected flow of customers that is crashing your server can be both an annoying problem (and I don't deny that you'll have to put a patch on it as quickly as possible) but also a concrete sign that you're moving forward. Your course is getting the success it deserves.

The real problem would never be dealing with situations like this because it would mean that you are not "pushing hard enough on the accelerator."

Make an Interim Evaluation. Every 6 months (or better yet, every 3 or even every month if you are a beginner), you take a step back to assess your progress. See if you are on track to reach your final goal and, if necessary, make corrections in the race.

You won't need a whole year to create an online course from scratch.

The world goes too fast and changes too quickly to hope that your plan is perfect and unchangeable.

Making changes must become not only a habit but a necessity.

The key here is not to give up completely or change your goal if you are not on the right track. Instead, you can make small changes to your course/goal, such as extending the deadline, removing some parts, splitting it into two sub-goals to make it more manageable and flexible.

Join a Mastermind Group. Mastermind groups

are groups of like-minded people who are hyper-focused on achieving their goals. Typically, a mastermind is composed of people who have achieved similar levels of success in different fields, so there is no direct competition but a fruitful exchange of information, experiences, and best practices.

In mastermind groups, there are conference calls (often via Skype/Zoom) or regular physical meetings where you can discuss your progress and, often, in each meeting, a different person will be on a "hot seat" and will have to talk specifically about its ups and downs, receiving feedback from everyone else.

Mastermind groups tend to take a long time. There is a high level of commitment; some even have an annual membership cost of more than 10,000 dollars, but the value you receive is much higher and pays back all the (economic and time) investments made. To teach subjects in which you are competent.

This should be obvious, but maybe it's better to put it into words since there's some much fluff around: you don't teach something you're not experienced in.

Creating An Outline

If you've completed the action steps for each chapter, you now have a simple sales page and know how you want to deliver your course. You may even have your first paying students. With this foundation in place, you're ready to start creating your course content.

An outline defines your course scope so that you can create it more quickly. During this stage, you decide what information you want to include and what you want to leave out. A strong outline also ensures you create the best experience for your students. By considering the individual journeys of your students, you can decide how best to deliver the promised solution to their problem.

There are four steps in the outlining process:

1. Define your start
2. Plan your end
3. Brain dump
4. Chart your path

This step doesn't have to take long. You can outline a short course in an hour or two. Larger courses may take longer, but there's no reason why

Define Your Start

Are you targeting beginners with your online course or more advanced consumers? You need to know where your readers are starting before you can decide how far to take them.

Write down a few sentences about where your ideal student is and what they need to learn next. Here are three examples to help you with this step:

My students already have their own blogs, but they're having trouble growing traffic and generating income.

My students are trying to work from home, and they have children. They have no idea where to find the time, but they want to start a business.

My students want to plant a garden this year, but they don't know what zone they're in, how to prepare their soil for maximum yields, or what plants are good for beginners.

Notice how these hypothetical examples narrow down your focus in just two sentences. You can't design a solution for everyone. Instead, pick your ideal audience's starting point, and go from there.

Of course, people of different abilities will sign up for your course. That's normal since no two people are alike. However, you will have a much more effective solution if you have a clear starting point in mind instead of trying to make everyone happy.

Plan Your End

How far are you going to take your students? What specific things will they learn by the time they reach the end of your course? What result do they most desire, based on where they are now?

Write down a couple of sentences about the outcome your readers can expect from taking your course. Here are three examples (which complete the starting points above) to help you with this step:

My students will select one of five methods to monetize their blogs and earn their first $1,000. They will implement four proven traffic-generating strategies to get more eyes on their content.

My students will create a weekly schedule and a meal plan to minimize their decisions. They will define tasks to work on when they have a few minutes to spend on their businesses. They'll implement five practical strategies to help them make time for their business as busy parents.

My students will prepare the soil for planting. They will learn how to read zone maps and identify their zone. They will compare and contrast popular plants for a garden. Finally, they will understand space requirements, how much work each one takes, and what to consider before planting.

Notice how the endpoints of each example are specific and share what the student will gain. Also, many results are quantified with numbers. For example, students will earn $1,000 or implement five practical strategies.

Brain Dump

Now that you have a start and endpoint let's see what else is hiding in that mind of yours. You need to do a brain dump to get all your ideas for your course on paper.

There are many brainstorming techniques. One popular technique is mind mapping. A mind map is a big circle in the center of the page with lots of connecting circles in which you'll write down ideas.

The following is an overview of the mind mapping process. Feel free to adapt it to how your brain works. The beauty of mind mapping is that it frees your brain to create. So, don't be rigid in following each step. Do what works for you.

Start by deciding what tool you want to use for your mind map. For example, if you think best with a pen in your hand, you may prefer pen and paper. You can also use Post-It notes by writing each idea on a separate note or free software tool (such as XMind or Coggle).

Next, draw your course topic in a central circle. Starting in the center gives your Brain freedom to spread out in all directions. Let your thoughts flow!

Begin writing your ideas around the central circle. These ideas are called branches. Don't censor yourself; you can always delete ideas later.

Connect your main branches to the central circle and connect your second-and third-level branches to the first and second levels, and so on. Your brain works by association. It likes to link two (or three or four) things together. By connecting the branches, you'll understand and remember a lot more easily.

Keep adding ideas until you run out of steam. Work on your mind map over several sessions. This allows your subconscious to come up with new ideas that you can add to your mind map. Write as much as possible. Cutting items is a lot easier than generating new ones, so keep going. This free-writing ensures you create the best possible course, so don't skip it. As you brainstorm, don't analyze your ideas too much. At this point, you shouldn't be rearranging them or figuring out the best order. Your goal is to get all your thoughts out of your head.

Chart Your Path

Now you have everything you need to outline your course successfully. You know where you're starting and stopping, and you have a mass of ideas for the main content. It's time to organize everything into a logical order. This is the course outline that shows the journey your students will take from beginning to end.

You can't magically move your students from where they are now to where they want to be. You must guide them along the journey through a series of milestones. This improves their journey by giving them a sense of progression as they take your course.

Sometimes, there are multiple ways to organize your ideas, and you can agonize over finding the best outline. Try not to let your inner perfectionist prevent you from finalizing your course plan.

Here are the steps to creating an outline from your mind map:

First, group related ideas together. You can do this by numbering or color-coding ideas on your mind map. If you're using Post-It notes, you can move them around on a big wall or floor. Your idea groups are your course lessons.

As you group ideas, you may notice some of them don't seem to fit anywhere. These ideas need to be scrapped. Ask yourself whether they are useful in achieving the promised outcome of your course. If they aren't, leave them out of your outline. You can always include them in your next course! Be ruthless when cutting elements. You need to deliver the promised result for your students in as few steps as possible.

Once you have your ideas grouped, put them into a sequence. Think about the journey you want your students to have. What information do they need to learn first? What steps should they take, and in what order, to achieve the promised outcome? You may need to try several outline ideas before you find one that flows.

For a longer course, you can also group lessons into modules.

As you complete this step, keep in mind any tests or assignments your students will have. Are you going to ask them to check in with you after certain lessons? Will there be any quizzes they have to complete before moving on?

When you create a course, you need to think through the student's experience. Making sure students understand what they're learning is an important component. As you work on your outline, consider adding a symbol next to any step that needs an assignment, test, or quiz. A simple star or circle will do. Just make sure it's something unique. Then, when you get to the creation stage, you won't forget where you wanted these special components, and you can prepare them as you go.

Once you have your outline, take a step back and look at it from the perspective of your ideal student. Is there anything that doesn't need to be there? Is anything missing? Do you see any repetitious spots that could be cut? Make changes as necessary.

Now, before you move on to the next chapter, there's one more important point about outlines. Ready?

Just because you create an outline doesn't mean that's exactly how your course will end up. In other words, your outline isn't carved in stone. As you dive into content creation and take your first group of paying students through your course, you may discover gaps or double entries. Be open to adjusting as you go.

Do a brain dump of all the ideas you want to include in your course.

Finally, organize your ideas into an outline by charting your students' path from the starting point to the promised outcome.

Choosing A Topic

Are you excited? This is going to be a fun and challenging ride. By the end of your journey, you will have learned new skills, created a product you are proud of, and earned your first dollars as a course creator. We will start by selecting a topic for your new product.

What could you teach others? If you can solve a problem (or a series of problems) for your audience, and are comfortable sharing your knowledge, you can create a course.

But there are a few things you must do before you start building. Don't skip this chapter, even if you already know your course topic. Otherwise, you may discover that what you design isn't something people want to pay for.

Your Ideal Student

You do not want to waste time building a product that doesn't connect you with your audience. Your future students must be at the forefront of every decision you make. So, take a few moments and think about your ideal student.

If you don't yet have an email list or social media following, that's okay.

Throughout this book, you will learn how to build a loyal tribe. Even if you don't yet have an audience, you must understand who will buy your course.

The following four questions guide you through this process. The first two are generic, designed to help you nail down who you want to help. The second two questions are ones you can ask for each potential course topic.

You can answer these questions in several ways. The first method is to do online research. Read relevant forums, online groups, Q&A sites, and reviews (Amazon, Udemy, etc.). Be on the lookout for discussions around actual products and services within your topic categories that people are buying.

The second method is to talk directly to your target audience. Meet people for coffee and ask them about their lives and challenges. Or, if you can't meet in person, chat with them online or on the phone.

Finally, you can create a survey using a tool like Survey Monkey or Typeform. Then ask people in your target audience to complete the survey. Circulate your survey via email, in a blog post, and on your social media channels. Aim for at least ten responses so that you can gather as much insight as possible.

Question 1: Who is your ideal student?

To position your course and make a profit, you need to know some basic demographics about your ideal student. Answer these questions to help narrow down your audience:

How old are they? What gender are they?

What is their marital status? Do they have any kids? What ages are they?

What are their common professions?

How do they consume content?

Are they looking for a quick fix or a deep dive?

How do they like to learn? Do they prefer text, audio, or video content?

How much time do they have to take a course?

Do they prefer formal or conversational style?

What books/blogs/media do they enjoy? (This can help you think about the tone of your course.)

Question 2: How does this person think?

Demographics aren't the only similarities your audience has. They also share psychographics. These are points of data that indicate how people think. Answering these questions can help you get further into the mind of your ideal student:

What limiting beliefs or fears are they facing?

What are their interests? What opinions do they hold?

What do they care deeply about?

What preexisting feelings do they have about the topics you're considering?

You may not immediately know the answers to these questions. If possible, take time to connect with people in your target audience in person. Don't guess or assume you already know how your ideal student thinks.

Question 3: What problems does this person have?

Now that you know a little more about your ideal student, dive deeper into their struggles. What keeps this person up at night? What would they happily pay someone to "fix" for them? These are your potential course topics.

Note that people are best at answering two types of questions: questions about past behavior and questions requiring them to voice an opinion on something.

Here are some questions you can ask for each problem your ideal student is experiencing:

Can you tell me about the last time you experienced [the problem]?

How did it make you feel?

What specifically about [the problem] keeps you awake at night?

What, if anything, have you done to solve that problem?

What do you like or dislike about the solutions you've tried?

Many successful courses solve a problem the creator personally experienced and overcame. This is the basis of both the authors' businesses. For example, Sally worked for years to build a business to stay home with her kids and make money without feeling torn between the two. She understands her audience because she's been through their struggles, shared their dreams, and still does.

Having a deep empathy for the people you serve gives you greater satisfaction in your work and attracts more of the right kind of people to your tribe. Now, you don't have to solve a problem you've personally experienced. But you must have a thorough understanding of your ideal students.

Question 4: Can you offer real solutions?

Make sure you can provide a solution to whatever problem you pick. Write down the course ideas that you believe can help your audience. Then, take a few minutes and brainstorm the solutions you can offer.

Don't limit yourself to solving complex problems. Many successful courses deal with bite-sized problems that are nevertheless urgent to the target audience.

Also, don't be afraid to add something that you think is common knowledge to your list. If something is easy for you, it doesn't mean it's straightforward for everyone else.

Once you have your ideas written down, ask yourself this question for each: Can I provide a real solution?

If the answer is no, cross it off your list.

Your Strengths and Interests

Once you have your target audience in mind, you can further narrow down your course ideas. To help you come up with the best ideas, review your strengths and interests. Building a course around a topic you care about and are comfortable with is much easier than working on something that doesn't interest you.

If you pick a subject you don't fully understand; you will struggle. It will take more time, and you will need to do extra research. It will be harder, which will make you question your ability to deliver a solution. Also, your audience will not connect with your marketing. They will be able to tell that your heart isn't in it.

Products you are passionate about are much easier to sell. So, make sure you align your interests and strengths with the topics your students want to learn. If you're not sure what your strengths are, answer these questions:

What comes naturally to you that other people find difficult?

What do people ask you to help them with?

If you were asked to present at a live event, which topic would you feel most comfortable talking about?

What is your educational background?

What types of jobs do you have experience with?

Are there any unique situations in your life that deeply impacted you?

Now, delete any course topics that don't match your strengths and interests.

Do a Money Check

Up until this point, we've focused on you and your audience. What does your audience want that you want to provide? Before you go any further, you must also check whether there's potential to make money from your idea.

You need to do a little more research to establish whether there's a market for each of the course topics remaining on your list. This research should only take five to ten minutes per topic. Your goal is to answer this question:

Are people already making money selling courses on this subject?

One way to do this is by looking at Udemy. Udemy is a platform where you can find online courses in just about any subject. To research your idea on Udemy, search for courses on your topic and see if they are selling. You can't see how much the students paid (they may have also received the course for free). Look for courses with many enrollments and ensure they have a range of reviews spread over different date periods.

A second way to see if your course idea will sell is to view the websites for bloggers in your niche. Can you tell which bloggers are successfully selling learning solutions in this space? Some bloggers even share reports of their earnings and break their income down by source.

As you do your research, eliminate any topics that have limited earning potential.

Now that you spent time brainstorming and analyzing, you must pick one topic and move on.

You may feel uneasy about committing to one idea. That's normal. Trust that whatever you choose now will be the perfect decision. Then never look back. You can't continue without a solid topic idea.

Defining Your MVC

Imagine creating your course knowing exactly what to put into your product and having absolute certainty that people will buy it. In fact, imagine if people purchased your course before you started building it.

Well, that's exactly what you will do in the next two chapters. You will find people who want to buy your course right now. You're not about to spend the next six months designing a polished product, only to discover that nobody wants it. You're going to sell the promise of a solution first.

You will only create your course after people have raised their hands and asked to purchase it. If you follow the process in this book, students will be lining up to buy from you. And better yet, your product will be of higher quality. You'll take feedback from your students to create the ideal solution for your target audience.

So, are you ready to test your idea in the real world?

What Is an MVC?

Your goal is to quickly test your course idea by asking people to buy it. And to do that, you need to describe your solution. In the action steps, you will define a Minimum Viable Course (MVC). This is the smallest offering that will serve your target audience. Let's look at an example.

Say your audience consists of moms with young children, and your idea is to build a course that helps them look and feel the way they did before they had kids. You know that new moms need a flexible solution. They don't want to spend hours at the gym.

Your MVC might be a six-week workout plan for busy moms. You could deliver lessons via pre-recorded video, live online training, or even email. We discuss course format and delivery options in chapter five. So, don't worry about the technical details right now. Instead, focus on describing the result you want to provide to your students.

Your MVC should fulfill the following criteria:

There's demand for the course, and you want to create it. You should already have proven this in the previous chapter.

You have the resources and knowledge to provide the solution.

You can create a minimum version of your course in less than two weeks.

Some people get stuck at the idea of creating something in less than two weeks. But you shouldn't build a complex course at this stage. You must move through the testing phase as quickly as possible. Don't spend months designing something only to discover nobody will buy it.

You can even develop your content as people are taking your course. In the six-week workout example, you might create the first week's training now. Then produce the remaining lessons over the next five weeks, dripping the content out one week at a time. This way, you always stay at least one week ahead of your students. There are other ways to create course content without compromising on quality quickly. For example, instead of pre-recording a video course, you could deliver training via live workshops. This needless preparation and your students benefit from having live access to you.

Remember, your MVC is the first version that you intend to improve in the future. You are asking people to buy the beta version of your product at a discounted price. They will receive additional support from you and access to the final version of your course once it's created. This is a win-win. You prove your course idea and make some money upfront.

Your students get early and exclusive access to the expert (you).

Finally, if the product you want to sell is too big to create quickly, you can still test your idea by pre-selling or pre-launching it. Both these terms are described in the next chapter. So, if you have a course idea but you can't create it in less than two weeks, don't let this stop you from moving forward.

Write a Sales Letter

Once you select an MVC, you need to write a sales letter for it. This exercise helps you refine your offering by clearly communicating the value to your potential students. Doing this *before* developing your content also ensures you create something people want and are likely to buy from you.

Your goal here is to create a no-brainer offer. Something so good you know it will be irresistible to your target audience. An offer you are confident (even excited!) to share.

An easy way to do this is by using the *60-Minute Naked Truth Salesletter Formula* by Dean Jackson. This exercise works well because it pushes you to get your ideas down on paper fast.

Below are the steps to write your sales letter. As you write your letter, hold a clear image of your ideal student in your head. If you know someone in real life who fits your customer profile, think about that person. The best marketing happens when you care about your students and understand their needs at a deep level.

Start with the purpose of your letter. "I'm writing to you because I want you to..." Insert the naked-truth reason you're writing, as if you're making your request known to a lamp Genie who could grant your wish, like, "Take out your credit card and pay me $147 for my new course called..."

Provide reasons you are writing to this specific person. "The reason I'm writing to you specifically is that I think you want..." And then, list the reasons in bullet form, such as reason #1, reason #2, reason #3, and so on.

List the features and benefits of your product or offer. "Here is a list of what you get when you [buy my course]...." Again, use bullets. First, list the feature followed by the benefit, such as, "You get [feature], which means [benefit]." Write as many as you can drum up at this point.

List top 10 questions and/or objections. You can say, "If I were to guess the top 10 questions or objections you would have about buying my course today, they would be..." Follow that by another bulleted list of the top 10 most asked questions or most pressing concerns.

Provide answers to those questions or objections. "So here's how I would clear those up for you..." Same idea as point #4. List, in bullet form, the answers to each and every question or concern you've uncovered.

Explain the guarantee or how you are removing the risks. "I want you to be completely without risk, so here's my guarantee..." Then explain how your guarantee works, how it reduces or removes the risk from the purchase in their minds, and how to take advantage of it if they need to.

The most important part is the call to action. "It's really easy to get started. You just..." (whatever it is they must do, such as "click this button," "fill in this form," "call this phone number," "return this coupon," etc.) Provide the exact, step-by-step instructions on how they can take action.

The result of following the call to action. "Once you decide to get started, here's what will happen..." Describe what's going to happen once they begin. Educate them on how they will get their product and how they will consume it. Tell them how to make the best use of their new purchase.

Add an element of scarcity or a sense of urgency. But only if this is REAL. Don't use false scarcity tactics (for example, saying the offer is expiring when it is not). "You need to do this right now because..." Tell them why they need to take action today. Is there a limit or a deadline? What will be the consequences if they don't take action? What are the ultimate costs of not moving forward today?

Finally, include testimonials from satisfied customers. "Here's a list of people who have already [done this] and exactly what happened for them..." Add testimonials or case studies from other customers. Of course, these must be real and genuine.

At this stage, you probably won't have testimonials. However, you may have resulted from the life that you can share. For example, if you've personally overcome the problem your course solves, you can talk about your own results instead.

If you're struggling to write your sales letter, review some sales pages for other courses in your niche. You can also see an example sales letter in the bonus area here: sallyannmiller.com/courses.

Don't stress too much over this step. Do your best and move on. In chapter eleven, you will learn more about how to create a sales page for your course.

Price Your Course

Now, you need to set a price for your course. Below is a simple process you can use to price any new product. It's quick and should help you avoid pricing paralysis.

At this stage, you're checking to see whether there's a market for your course. You want to set a "no brainer" price. This doesn't mean you will compete on price over the long term. But for now, you need to find out whether people will pay for your idea.

Start by imagining that you are your ideal student. From the prospective of your ideal customer, eager for the solution you provide, ask yourself:

What is it worth to have this problem solved?

Answer with the first number that comes to mind and writes it down. Don't overthink this step or waste time worrying about what other people charge. Simply focus on selecting a price that's consistent with the transformation your course offers. Also, make sure the price you write down is a little outside your comfort zone.

Now, make sure your price is realistic. When testing an idea, it's important to set a price that will allow you to find your first paying customers quickly. Don't worry; when you've gained experience and testimonials, you can increase your prices.

Review the number you wrote down in light of the following:

1. Your course topic (what's normal in your niche)

2. Your background and experience

3. The transformation your course delivers

4. What's included in your offer

Does the number feel realistic? Will your ideal student pay this price to take your course? If necessary, lower your price. But make sure your final number still makes you a tiny bit nervous. When someone pays you that magic number, you'll push yourself to provide massive results. And massive results make for happy students, glowing testimonials, and repeat customers.

This new number is your target price. It is not your no-brainer price. This is the price you eventually plan to set for your course.

To turn your target price into a no-brainer price, you want to discount it by anywhere between 50 and 75 percent.

Action Steps

1. Follow the steps at the start of this chapter to define your MVC. Make sure your idea meets the following criteria:

There's demand for the solution, and you want to create it.

You have the resources and knowledge to provide the solution.

You can create a minimum version of the solution in less than two weeks.

Now, refine your course idea by clearly communicating the value to your potential students. The goal of this activity is to create a written description you can share with people. Follow the steps described earlier in this chapter and write your Naked Truth Sales Letter.

Finally, refer to Price Your Course above and select a no-brainer price for your test offering.

You have now completed the first two steps in the eleven-step system. You have selected a course topic and created your MVC. In the next chapter, you will test your idea by recruiting some people to take your course. Get ready to make your first few dollars from your course idea!

Testing And Evaluating Your Ideas

People who reject your idea are not life-threatening. In fact, it's not even a failure. View this next step as an experiment. If people don't take you up on your offer, that's okay. This is feedback and nothing more.

Some people want to learn what you can teach them. It's your job to find those people and figure out how to communicate your idea to them. So, know that any fear or doubt is normal. It means you have a human brain. We all experience self-doubt. The people who succeed are those who press onward and move through fear.

Create a Simple Sales Page

In chapter eleven, you will learn more about what goes into a compelling-

ling sales page. For now, it's important to keep things simple. All you need is a place to share the sales letter you wrote in the previous chapter and a means to collect payment from clients.

There are many online options to help you create a sales page, such as LeadPages, SamCart, ClickFunnels, and Kajabi. These tools provide templates for your sales page and typically integrate with a payment processor. Also, many learning management systems have built-in sales pages and a means to collect payment. We discuss some of these platform options in the next chapter. However, all these solutions have a price tag associated with them, and some are easier to set up.

If you're not comfortable learning new technology or don't want to make an investment yet, you can create a simple page on your website. Copy your sales letter onto a new page and add a PayPal button. If this feels too complex, you could even write a social media post. Invite people to email you if they are interested in buying your course and invoice them directly.

Select whichever option is easiest for you to implement right now. Don't let technology slow you down. Your goal is to share your course idea and see if anyone wants to buy it. Once you've picked a platform, build your basic sales page. Then move on. Don't worry about making it perfect.

Recruit Students

Your first group of paying students will go through your MVC and provide feedback about the content and delivery. Selling a beta version of your course is the fastest way to test your idea. You can also use the information from your early students to refine your course before launching it to the masses.

Most of us don't want to risk rejection. This is normal. As humans, we've evolved to avoid rejection at all costs.

Thousands of years ago, being cast out of the tribe meant death. The primitive part of your brain works hard to help you avoid any form of failure. We recommend asking your students to pay for your course instead of offering it for free. Purchasing a course (even at a no-brainer price) gives people "skin in the game." They're more likely to work through it since they paid for it. And, of course, charging people tells you whether there's a market for the solution you're offering.

Payment isn't the only detail you need to figure out. You also need to decide:

- How long the beta process will take.
- What you expect in return.
- How long will students have access to the material?
- What will you do with any information and feedback they provide?

You want to have this process clear in your mind to share it with others easily. Make sure your students know in advance what they are getting and what you expect from them.

Once you know what your students will be doing, you can start recruiting some. However, it's important to remember that you want members of your target audience to take your course. Don't try to sell your course to just anyone. Look for your audience in places where you know they hang out.

Use what you know about your ideal student and present the opportunity in places where they will see it. Use language they'd appreciate, and explain what pain points your course covers. Help them understand whether it's a good fit for them.

You want to have a small group of students to provide a variety of insight and establish whether enough people will purchase your course. There isn't a "perfect" number, so aim for three to ten people. This will prove there's a market for your idea and give you enough feedback to refine your solution.

Below are five places you can share your course. But don't feel constrained by these. Use your imagination and be prepared to keep testing different ideas. We go deeper into building a tribe in chapter nine. For now, your goal is to get your course in front of enough people who will find it relevant enough to purchase.

When sharing your course, you want to start by giving free value. This builds trust and demonstrates your expertise. Focus on serving people. Let go of your attachment to selling your course and commit to helping the people who need your solution. You'll soon discover that what you put out into the world eventually comes back to you.

Here are five places you can show up and give value to people interested in your course topic:

If you already have an email list, start here. Send helpful emails relevant to your course topic.

Participate in forums and Facebook Groups related to your topic. Answer questions and be a helpful and active member of the group.

Publish helpful posts on your social media profiles. Provide insight into your topic and share your journey with people who follow you.

Attend in-person networking groups and connect with people you know in real life. Look for ways you can give value and be of service to others in your local community.

Post-long-form articles about your topic on Medium. This is a free platform that has a built-in audience. Now, you may be wondering when and how to tell people about your course. After all, the goal here is to discover whether people will pay for your product and recruit some students.

Make offers as often as you can, but do so from a place of service and not from neediness or desperation. In other words, don't cold call people on their personal number or send them private messages asking them to buy your course. Instead, consistently share your basic sales page in the places where you're already giving value.

If you have an email list, then you can be more aggressive with your approach. These are the people who have already opted in to hear from you. They want to know about anything new you have to offer.

Remember, being aggressive means emailing more frequently. It does not mean pushing your course on people. Never be desperate to make a sale. Aim to send at least five emails announcing your MVC and why people should get in early and become beta testers.

Here's one final tip that will help you sell anything. Imagine you have the cure for cancer. You wouldn't send pleading messages to people who may or may not need your help. You would shout about your cure from the rooftops. You would announce it to the world. Then you would wait for the people to come to you.

There's no need to obsess over how often you share or the exact words you use. Instead, believe your course is valuable and can help people. Do this, and customers will come to you without you having to chase after them.

Pre-Selling and Pre-Launching

In the previous chapter, we mentioned pre-selling or pre-launching your course. These are both ways to test your idea if you can't create an MVC in two weeks. Now, let's spend a few moments describing each of these options.

Pre-selling is where you ask people to pay for your offer before you create it. This is the concept behind crowdfunding sites like Kickstarter and Indiegogo. Writers and creators can also use Patreon to find supporters who will pay for their creations. A previous client of mine even pre-sold her children's book with only a rough draft in her hand.

Suppose you are pre-selling a course that won't be available for several months. In that case, you will need to offer some extra incentives to encourage people to invest now. For example, extra input into the course creation process, additional coaching with you, or access to an in-person event. Your only limit when it comes to bonuses and incentives is your imagination. We talk more about offering course bonuses in chapter eight.

Pre-launching is where you collect email addresses from people interested in your idea. It allows you to build an audience while you build your course. It's a valid strategy but less effective than collecting money for your idea. People give away their email address more easily than their cash.

By having a waitlist, you can see how many people are interested in your course and get early feedback on what to include or leave out. You can create a waitlist between one and three months before your launch. Just make sure you keep in contact with people so they don't lose interest before you launch.

What if Nobody Buys?

Building an online business is an iterative process. Few entrepreneurs find success right out of the gate. If nobody wants to pay for your MVC, don't panic. This is part of the process. It's data and nothing more. You test your idea, analyze your results, and then look for ways to improve. If you're prepared to keep testing and improving, then success is inevitable. If nobody is buying your MVC, take some time to find out why. Survey people who saw your offer and didn't buy. Here are three questions you can ask:

Were you interested in buying the course? In your own words, why didn't you join?

What will you do instead to solve this problem?

Answers to these questions tell you whether your course topic is interesting and/or whether you are doing a good job of presenting the value of the transformation. Look for patterns behind what respondents tell you. If people say they don't have enough time or money, dig deeper. These are often excuses that people give when they are scared to make a purchase. Address any specific reasons by changing your offer and sales letter. You will want to remove any doubts your potential students may have.

Occasionally you may need to go back to the drawing board and pick a completely different course topic. But at least you've found this out before you have built a product. Take the information you learn, and try again. The only way you will fail is if you give up.

In the action steps, you will make a commitment to show up and help people every day. Keep putting value out into the world and sharing your course idea. Do this until you either have some paying customers or realize it's time to test a different idea. But don't keep taking the same action if it isn't working. You must tweak your offer and what you're sharing based on feedback. This stage is all about discovering what you want to do that person will also pay you for.

Action Steps

Create a simple sales page for your MVC.

Make a commitment to show up every day in at least one of the places described above in "Recruit Students." Write out your commitment and share it with someone—an accountability partner, your spouse, or a mastermind group.

Start showing up and putting value out into the world. And tell people about your course—each time you share or connect with a person in real life, check-in with yourself. Make sure you are coming from a place of service and not form an attachment to making the sale.

Keep executing your plan until you either have paying customers or decide to test a different idea. Make sure you keep tweaking your offer and what you're sharing based on feedback.

This is the hardest step in the process, and it's normal to encounter some resistance at this point. In most cases, your hesitations are related to fear and uncertainty. It's scary to put yourself out there. Sharing your ideas and asking for money is not always easy. Please trust in this process.

Keep testing and be open to failure. If you're willing to fail as many times as it takes, then success is guaranteed.

Congratulations. You made it through the hardest step in the process. This takes courage, patience, and resilience. Make sure you celebrate your achievements, no matter what the results. You showed up, did the work, and faced your fears. In the next chapter, you will pick a platform on which to deliver your MVC.

Choosing The Right Platform

Have you created a simple sales page and started telling people about your course?

Now that you're on your way to having your first paying students, it's time to create the initial version of your course. We start by deciding how to present the information in your course.

You have two key decisions to make. The first is the format of your content: text, video, or audio. The second is how to host your course content.

The decisions you make now aren't fixed in stone. So, for your MVC, stick with a format and technology you're comfortable with. Your goal is to create your MVC in two weeks or less.

Once you've proven, there's a market for your idea and gathered feedback from your beta testers. You can refine your course content and delivery.

Select Your Course Format

Let's start by deciding on a course format. The three common formats

are text, video, and audio. All of them have different pros and cons. ***Text-Based Lessons***

Text-based lessons are simple to create and easy to maintain. You don't need new technology, and it's quick to make updates in the future.

Written lessons also offer flexibility. Students can skim through lessons and change their reading speed based on their situation. Moms can read text-based lessons while snuggling with a baby during naptime without worrying about waking their little one. The lack of sound also makes this type ideal for completing at night or waiting in a line.

However, text-based lessons are not the most concise. The written word isn't the easiest way to explain some complex ideas. Many students learn better through watching or listening to course content.

Finally, text courses have lower perceived value when compared to other formats, so this may impact how much you can charge for your course.

Video Lessons

Video lessons can be pre-recorded, or, in the case of your MVC, you can record yourself teaching the life lessons.

Video lessons are perceived as more valuable. Many high-end courses are video-based. Society values this type of content currently, and many people prefer watching content to reading it.

You can also create videos quickly and present ideas by showing rather than telling. This is especially helpful if you're teaching something that requires visual aids, such as technical instruction. Finally, your personality shines through in the video. You get to connect with your students in a more authentic way than you can use text material.

There are disadvantages to video. It requires some technical skill and equipment. For example, you will need a decent camera and recording software. To make your videos look professional, you also need good lighting. You may need to invest in a ring light or other portable lighting options.

When you sit down to write a lesson, it doesn't matter what the background is around you. No one is going to know there's a pile of laundry waiting to be folded on your table or clutter on your desk. But, if you record a video, your background becomes important. Your recording space doesn't need to be fancy, but it does need to be free from distractions.

Videos take data. For the majority of people in the United States, this is not a problem. However, those in rural communities may have limited data and may opt-out of purchasing your course, knowing it'll drain their metered supply.

While the video is quick to create, you should also consider accessibility. For instance, someone who is hard of hearing will not be able to listen to your videos. For the final version of your course, you may decide to provide a written transcript. This adds an extra step to the process.

Finally, fear holds some people back from creating video content, as does the quest for perfection. You must overcome these mental obstacles to teach through video successfully.

Audio Lessons

Audio lessons offer a different set of advantages and disadvantages. They are simple to capture and require less technology than video.

With a decent voice recorder and microphone, you can quickly record your lessons.

You can create audio content as if you're speaking to a friend. This conversational tone is perfect for sharing information with others. Also, there's no need to worry about what you're wearing or what your background looks like, though you do need to pay attention to the noise level around you.

People often listen to audio content while doing other things, so they might not be as focused on the material. While this offers students increased flexibility, it can prevent people from learning key concepts.

As with video, you may want to create a written transcript so your course is accessible to people with special needs. This adds a step to the process.

Finally, there's a learning curve to creating quality audio content. You need to be comfortable speaking to help make your lessons effective and enjoyable.

Combination Courses

While some courses are entirely one format, many are a combination. You might find text lessons with some videos sprinkled in. Or video lessons with a written summary under each video.

You could even have an audio course with a few videos for when students would benefit from seeing a demonstration of some kind. Your imagination is the only limit when it comes to creating courses. You can mix and match as you see fit.

However, please do not randomly decide on a format. Pick the one that makes the most sense for you and your audience. Also, remember that the format you pick for your MVC isn't necessarily the same format you use for the final version of your course.

Hosting and Delivery

Your second decision is how to host and deliver your course. You have many options. Some cost more but require less technical know-how. Others are quite inexpensive but require you to handle any technical problems that arise.

Examining every platform available is beyond the scope of this book. Instead, we're going to look at four types of hosting and offer some suggestions for each. Please do your own additional research and decide what method will work best for you.

As always, keep your audience in mind while you make this decision. You will want your course to be easy for them to access. You'll also want them to enjoy the experience and not dread logging into your course.

The four main options for hosting your course are:

- Email course
- Self-hosted Learning Management System (LMS)
- Course marketplace
- Hosted course platform

There's a fifth option that may be appropriate for your MVC. You can host your beta course inside a private Facebook Group. Record your lessons as live Facebook Videos or upload content you create in advance. This is a fast and free solution which I've seen used to good effect.

However, you don't want to host the final version of your course on Facebook. Social media platforms can change at any time. You don't own the data on Facebook, and you need to be careful that you don't fall foul of Facebook's terms and conditions.

Now, let's walk through each of the four main options.

Email Course

To create an email course, you need an account with an email service provider. Then you can set up a sequence of emails that are automatically given to your students over time.

This is the lowest cost option, especially if you're already paying for an email service provider. You can include a variety of lesson types by linking to video or audio files and attaching downloadable workbooks. Plus, students have access to the content without having to log into another site.

However, there are disadvantages. You need to integrate with a payment processor, such as PayPal so that people can purchase your course. You also need to integrate with your sales page. There are several moving parts to set up and maintain. You are responsible for all the technological components.

There's no simple way for you to see how much progress students are making. Some students might feel it's clunky or outdated.

Self-Hosted LMS

A Learning Management System (LMS) is software that allows students to interact with your course material. It tracks student progress, delivers the material, and provides you with report information.

A self-hosted LMS is installed on your own website, typically using a plugin. There are several plugins available for this. LearnDash, LifterLMS, WP Courseware, and Zippy Courses are a few popular options. They each have different costs, features, and benefits, so be sure to do research if you go with an LMS on your site.

Some benefits of using this method are low costs and increased control. Some plugins are free; others have a one-time or recurring fee. There's no middleman; you don't have to share your profits with anyone. A self-hosted solution also gives you flexible branding. You can ensure your course matches the rest of your site.

Of course, installing an LMS plugin on your site isn't all sunshine and roses. You're responsible for the technology, and it can be complex. There's a steep learning curve to set up your course and connect the payment portal.

Also, if you are handling payments on your site, you must handle security for your customers. How are you going to protect their private information?

Course Marketplace

There are marketplaces dedicated to hosting and selling online courses. You may decide that adding yours to one of them is the best move. Udemy is a popular platform in this space. Skillshare is another.

Marketplaces are easy to use. Most of these systems are intuitive, and you can set them up with step-by-step directions or tutorials. You're not responsible for the technology, collecting payments, or customer privacy.

The biggest advantage of a marketplace is that you get access to an existing group of course buyers. This may be desirable if you haven't yet started to build an audience of your own.

Online course marketplaces aren't for everyone. Here are some important points to consider before selecting this option.

You have no access to student emails, so you can't add your students to your email list. Also, there is typically a commission for this service, which means you give up a percentage of your profits. You don't get much say in how your course looks or is delivered. In some cases, you can't even set the price for your own course.

Hosted Course Platform

Finally, some people use an all-in-one LMS to deliver their course content. Podia, Teachable, Thinkific, and Kajabi are all options in this category.

Unlike course marketplaces, these platforms provide you with a bit more control. You get to set the pricing for your course and collect contact information about your students.

Students are familiar with these platforms. They are quick to set up and use. You don't need to integrate several different tools and are not responsible for student data or collecting payment.

Many all-in-one course platforms include more advanced features like affiliate programs or the option to sell a membership program.

While these solutions can be a great option, they are not for everyone. Typically they have a higher monthly fee for access. If you are just selling one course at this time, it may not be worth the cost.

The features may not be a great fit, either, and you may find yourself paying for features you don't need. You also have limited customization options.

Which One is Right for You?

Now that you know some of the pros and cons of each of the hosting options, did one of them stand out to you? Before finishing this chapter, there are a couple of important things for you to keep in mind.

You can change your mind later. If what you select doesn't work longterm, you can always choose something else. This means you can start with an "inexpensive" option and upgrade as you make more sales.

Many people get stuck in indecision. Make a choice and move forward with it.

There is no perfect solution. No matter what you choose, features of other systems may spark envy. For now, go with what meets your needs at a price point you are comfortable with.

Setting The Right Price

If you want to sell courses online, pay attention to how to calculate the selling price.

On the web, you can find any information of any nature, completely free of charge. We live in a time of plenty of information.

So even before asking the price to be attributed to a training course, it's worth asking yourself if it's not appropriate to abandon the idea of monetizing a course right away.

Doesn't this abundance of materials, resources, videos, articles make it completely useless to create a paid training course?

Free and Paid Training

Free training has totally replaced paid training. The answer is counterintuitive. In terms of volume, paid online training has been growing for years, and it is not decreasing. People pay, and pay generously when it is worth it— to enroll in a course.

Why? Because quality training still has room. There's room for everyone and also because free information still requires a research effort on the part of the student (and results are not guaranteed).

An online course built ad hoc by an expert is an accompaniment to competence if you think about it. Maybe one gets there anyway with free resources, but it could take years to reach the same level as an online course.

So there's room for both. Free information (easy to access but messy, of dubious origin, unstructured) and paid information (delivered by experts, well organized in the fruition, on a single platform).

Paid training courses exist, therefore, and are well sought after. All this brings us directly to the theme of this chapter, that is, what price to give to an online course.

Paying for a Service You Can Find for Free

The problem with online courses is that engagement is very low. Many students start a course and do not complete it.

This phenomenon affects us all when we are dealing with a good or a free service.

Think about it. We keep the products that cost a lot well, take care of them, and be careful not to break them. The products that cost little or even free will no longer consider them after a while.

It's the same for an online course. I don't know if you've ever paid for a course, but I bet you've attended a paid course very carefully, while with free courses you had no difficulty in leaving them halfway.

It's a paradox, but as a trainer, you have to think about it: if you really want your students to learn something through your course, it's good that you make them pay for it.

Your goal is basically to train people; your course exists solely and exclusively for this goal.

If you love your students, make them pay! **Low or High Price?**

A training course can be priced from free to a few thousand dollars. What about your course?

Here's my strategy: it is better to place a course at a medium-high price range, let's say not below $100.

In this way, you exclude students who are not ideal, those who do not understand the value of your course, those who would drop out after a while, those who are not really interested, and those who could easily ask you for a refund.

Now, only those people who believe in your education, believe in your person, believe that you have something important to convey would remain. It would be much more interesting to work with these students.

Selling a Benefit

I invite you to open your eyes to the most decisive issue in deciding the price.

With a training course, you sell a benefit, an accompaniment to resolving a problem, which a student of yours feels the need to solve.

This last step is important; I invite you to reread it and digest it in depth.

Your hypothetical student doesn't care how long the course is, if it has professional-quality videos, if it has well-designed PDFs, or if you are tall or short, man or woman.

What matters is that your course solves the need your student feels. If hypothetically, your course lasts only half an hour. Still, it solves an important and very heartfelt problem. It can have a market even with a high price (the concept of a high price should be outdated by now if you have followed it up to here: there is never a high price, just an inconsistent price).

The reasoning you should do is further upstream.

Make sure you design a course that has an audience of potentially interested parties. Don't make the mistake of creating an audience of potentially interested parties by lowering the price of your course immeasurably, because this is not a strategy, but a failure of a project, both from an economic and educational point of view.

If you have to lower the price to ridiculous levels to sell your course, something went wrong during the needs analysis. That's why the market reading phase is fundamental: try to sell something sought after by people.

Here you have four ways to increase the perceived value of your course and better justify the sale price:

Open the doors of your course for short periods. This strategy automatically increases the perceived value of your course because it creates an aura of exclusivity.

Provide various media. Some love watching videos, others love reading, others love listening. Allow all types of students to enjoy your course in the way they want to.

Offer custom counseling. Of course, a consultation can be offered if the price of admission is not too low (see the previous paragraph).

Keep a small group of students. You can follow them thoroughly and make sure they finish your course really happily (they will give you enthusiastic feedback, which you can then use in your promotions).

A Possible Strategy for New Courses (Or for Inexperienced Teachers)

In the case of a new course or an unknown teacher, it is good to design a price scale to be adjusted over time.

Let me explain it better.

At first, you won't have much feedback, and you won't have great testimonials; you won't have much movement on your sales pages.

What you need is the first group of students who can help you improve the course and sell it better.

In this specific context (the above about the low price is confirmed anyway), it is a good thing if you give discounted access to your course for a limited period, saying —among other things— that the price will go up.

It is the opposite of what we were saying earlier about lowering prices to have a flow of students. However, in this case, it is the opposite: you start low and then gradually increase until you reach the final price.

In this way, you ensure the first students, the first testimonials, and the first earnings. Maybe they may not be high, but it's a long-term strategy: you lose some money at the beginning and then consolidate your position in the medium-long term.

Do you understand how much you want to earn when you sell a course online?

Know-How Much You Want to Earn

In my opinion, this is another very interesting element when you have to sell courses online.

Ask yourself how much you want to earn from your training business. Is it a hobby that brings in some extra money a month? Is it an attempt to create a real online business?

Whatever your goal is, try to quantify it in a precise figure. Then you'll go back to figure out what price to give your course and how many copies you need to sell to reach that initial figure.

I warn you that math is coming (but this is interesting).

Final earnings = (contacts on list x conversion) x price.

Course price = final earnings / copies sold.

Let's give examples.

If you want to earn $5000 from your course's sales, just sell 100 copies for $50 each. Nothing so impossible. See how interesting math is?! This article aims to make you think about a couple of considerations about online course prices.

When you need to sell courses online, the best thing is always to take some tests and see what the right price is for your course.

In general, however, if it ends too low, it can't be the right price. Obviously, the problem is not in the price but in the previous analysis.

A good course must allow you to earn some money. Otherwise, the game is not worth the candle.

Marketing Your Course The Right Way

Can you take advantage of video courses to earn money online and create a profitable business?

If it is a passion, it is always useful to create the course, maybe not touching up every detail or technical aspect, but creating an asset helps the creator become good at creating video courses. It makes sense to do a market survey before promoting it and spending money on it, for example, checking on Google if there are already searches related to that topic (e.g., "how to sell on Amazon"). The minimum number of searches varies for each sector, but it shows who is looking for it, and even if few, they are very interesting contacts.

Is it a good sign if there are a lot of competitors? There must already be a market, with some online research. Then we would have to create further demand, for example, by going to a specific sector/service by promoting a specific product/service, when before there was only "generic" stuff. In general, it's better not to be a pioneer in a virgin market.

A good reason to create a video course (even if there is not immediately a profitable market) is that you learn the technique to make well-made video courses, and this will allow you in the future to be better and faster in creating and updating other products (or to create a service based on these skills).

Moreover, whoever does the course should be a trainer and a real expert who does what he teaches every day, to be updated and concrete (otherwise you will end up losing practical aspects).

Pro tip: The real business for experts/entrepreneurs should not be the course itself but the requests for a consultancy from those who have tried the course and prefer to be helped in achieving the desired objectives and

What to Do to Promote and Sell Your Course Online?

If you don't sell the course, it's difficult for the profits to come automatically, so something has to must at the marketing level (since the initial focus must remain to create a good product that makes people transform, otherwise marketing only "crap").

I suggest you create various courses on the most popular platforms (mainly social networks) and promote them through small introductory webinars of 45/50' followed by a promotional offer (about 60/70% discount) and that allows you to cover the costs of the lead generation and production of the course itself.

Once you have created a webinar that converts participants into clients, how do you bring people into the "virtual" classroom? Simply with two social networks: Linkedln (through direct invitations to entrepreneurs) and Facebook for large numbers.

As webinar platforms, I recommend Zoom, direct streaming on Facebook, GoToMeeting, GoToWebinar, WebinarJam, Google Meet, Stream-yard. There are many options, and I have tried almost all of them. The best one is Adobe Connect, but it's too complex, so, in the end, the choice fell on Zoom, stable, fast, and super reliable.

The Importance of Having Lots of Valuable End Products

It is a definition derived from the Hubbard Management System and means a valuable product or service that can be sold on a market in exchange for "sustenance" (money, goods, other services...).

It is not the realization of the course but the result/transformation/ great promise that the user wants to obtain: the video course is a tool to reach their goals. Therefore, it is not important that the course is bought, but that it is "done" so that the transfer of skills is completed (unfortunately, some statistics say that 97% of courses are never even started).

The responsibility is not only of the user but, above all, of the creators who, the more focused only on profit, the less attention he will devote to details. However, suppose you want the money to be the consequence. In that case, you'll have to do a cool and well-done course, not being able to count on exaggerated marketing and a big budget to spend on advertising.

Doing this is not difficult, but you need a strategy and a system; otherwise, it is not at all possible for people to start and finish the course.

It's not enough to create content to make users arrive because it's not automatic. Having the best product in the world is no guarantee of success: you need to make it known and promote it. There is a lot of bad information about it that leads people to say "courses don't work," but it doesn't: it doesn't work if it's not applied in the right way.

How to Promote and Sell a Course Online

Once you have finished your training product, the real challenge begins: you need to promote and sell a course online. To be more precise, everything could already begin in the production phase of the course. In fact, with a good promotional video spot, your course will certainly have more chances to go viral in a short time. But before starting, you have to take into account our reality today together with other objective factors.

Initially, I sold, like many info-marketers, info products like ebooks, guides, and my first online courses with the classic method:

- Create a squeeze page or sales page.
- Autoresponder and programmed follow-up.
- Free initial product in exchange for an email.

- Facebook ads to promote my landing page.
- Ongoing optimization of the automatic business.

My friend, this method has been outdated for a long time; I don't deny that even today, it works, but over time people have understood the mechanism. It has become inflated: in a few words, people got tired! What changes today? People are looking for valuable content. They are more careful when they want to invest their money in training programs. They want to be 100% sure to choose the right course, so you also need to change your inbound marketing and list-building strategies.

To promote an online course in a winning way (i.e., I'm talking about getting 300 students enrolled within the first two months of its launch), I have studied and created over the years an "Action Plan for the sale of courses" model, which is a six-step step-by-step guide helps to drive a ton of students to their course. This is a model that I continuously update because, as you see, some portals and tools can be used to promote courses depending on the niche. Now make yourself comfortable, have a coffee, and stay focused on what I'm about to explain.

Before we start!

To start promoting your online course, you must first consider the target group of your course:

Who is this course for?

What purchasing power does my target audience have? What are potentially other competing products on the network?

To answer these questions in an illustrative way, we will take an imaginary course in event organization as an example throughout the guide. So to answer the question above, we will say:

Who is this course for wedding planners, entertainment agencies, art directors, artists, event managers, or simply those who want to enter the industry?

What purchasing power has my target: in most cases, workers or freelancers with an average purchasing power ($99/$150).

What problem can people solve by buying my course: the course is focused on the marketing part of events, i.e., on a specific competence that no school, professional entertainment institute, or other training center teaches in this sector.

What are the potentially other competing products on the net: there are other courses.

Having this scheme in mind, now you already know how to address my potential customers to sell my course at best, so the first thing to do is make some arrangements on my sales page.

Step 1: "Settings"

The first step is essential to fix and adjust the appearance of your course before starting to promote it, to avoid, of course, burning your customers:

Creating the Sales Page. It consists of creating your own course page on the chosen platform, creating persuasive copywriting, a second introductory landing page, creating the covers, and all marketing and promotional material of the course.

Free Lessons. Showing a few lessons for free allows the potential customer to realize what s/he is going to buy. When you buy a course online, you do it because you want to trust the creator. You don't know the course's quality. The free lesson is a way to reassure the user about the purchase.

Create an Introduction Video. The introduction video of the course is a fundamental gimmick that can help you to summarize and make your potential customers understand in a short time what advantages the purchase of your course can have: I'm not talking about explaining what you will be dealing with, but about making people fall in love with a dream: in this specific case I used a more specific sales technique, i.e., during the promotion of my course in the classroom I gave the possibility to people who did not want to spend $500 to participate in enrolling at $149 to the online course: a double result advertising campaign. What promise did I make? I brazenly told people that thanks to my course, I would teach them how to earn $15.000 with a single event following a revenue model I had created. Do you think they called me a clown or a charlatan? Exactly! And there were so many of them! To make them understand that my method worked, I recorded video reviews of my students in that course. I used them in my next course as the main promotional tool. After all, if I were a blabbermouth, my course would have no more than 1,300 students.

Step 2: "Pre-Launch"

The machine is well waxed and ready to go. Now you need to organize the latest key activities before you massively promote your online course.

a. **Test Your Product.** Your course must work. I'm not talking about checking if a video is visible or if the final test answers are correct, but actually to have the product tested before launching it on the market. To do this, you can hire a freelancer on User Testing (a company specialized in website analysis, apps, and online services) and have video feedback of what works or doesn't work on your course. Obviously, an accurate analysis costs about $49, but if you don't want to make this initial investment, you can contact colleagues or other experts in the field of your course, ensuring free access to get feedback and in case you need to understand what to improve your content.

b. **Have Your Course Reviewed.** Your course will never have a single paying member if there are no reviews from other students who have already completed the course and have made their own judgment. Would you ever book a hotel that has no reviews? Here it is the same. To get your course reviewed, you have 3 ways you can follow:

1. Send free access to the course to your former students or your contacts, allowing them to attend your course for free in exchange for a truthful review (positive or negative, that is).

2. Find forums or specific industry groups: I am a member of several groups on Facebook and Linkedin of event organization and entertainment, and for each group, I have provided 10 free coupons for my course. Within 2 weeks, I got 5 reviews from my first students.

3. Invite industry experts to review your course; what better than a famous person to review your course? You can take advantage of your knowledge in a positive way to have significant benefits by asking influencers in your field to review your course. Ask only those you know directly! Here's a template "fill in the blanks" that you can edit and use as an email to send to an expert to ask for a review:

"Hi,
I am writing to you because I'm finally about to launch the new online course experience I have gained in my field also comes from your suggestions that I have read and followed through your articles/ video. That's why I thought it was right to reciprocate by giving you a preview of the exclusive access to my course for free; I would like to have sincere feedback from you on the course to understand how to improve my product eventually. I'm waiting for your answer. I wish you good work!"

c. **Start a Countdown.** Start by letting your audience know that you are about to launch a new product. If you have a Youtube channel, launch a short video where you tell them that you are about to preview an online course. On your blog, social channels, you can always share these videos and also write a few posts where you tell the audience that you are working on a new product.

d. **Send a Preview Offer to Your Contact List.** Send a preview newsletter to your database of contacts where you explain the advantages of your course in five simple lines, giving them the possibility to have a 70% discount on the sale price of the course as your personal contacts (the offer to stimulate the rush should not last more than 3 days). If you don't have an email database, you can create it in 4 steps: Login on Linkedin > Go to links > Select all and click on Export > Import the CSV of your contacts on Mailchimp. Now that you have your email database, composed by your professional links, to do it alternatively with Facebook contacts, you just need to create an email account on Yahoo, synchronize your contacts by logging in with Facebook, and then export a CSV with name and email address of your Facebook contacts (be careful here we are a bit on the limit).

e. **Invite Your Trusted Contacts to Support You.** Here it's time to dust off years and years of networking with colleagues in your industry. Invite your best contacts to report your course as a great new training program. This will allow you to attract the attention of people targeted with your training product and most likely be willing to provide their contacts to stay updated on the launch date of your course.

Step 3: "Launch"

Finally, the moment of launch has arrived, now what you will have to do is concentrate on finishing your promotional strategy. Before investing, I advise you to follow other steps for the organic promotion of your course:

a. **Use Training Course Directories.** About 50 well-indexed directories in the training world can help you carry tons of traffic to your online course.

b. **Use Sector-Specific Training Directories.** In addition to general training directories, there are several niche directories for certain professions.

For example, it is very easy to find websites specialized in ads for courses aimed at psychologists, lawyers, accountants rather than cooks, hoteliers, and so on. These portals allow you to get more qualified traffic than normal directories; I advise you not to insert a link to your course in a portal off-topic because it may not be successful as a strategy, or even you could be banned as off-topic.

c. **Use Industry Portals.** I discovered, during the promotion of several online courses, the great strength of the industry communities. I'm not joking; when I started promoting our course on Bach flower treatments within websites focused on this niche, I was amazed by the results obtained. About 3000 visitors in 1 month and over 100 sales with a conversion rate of 3.4% on that specific course and 5.12% on other similar courses. I admit this strategy kept me busy for about 5 hours, but it is worth it with these results!

Step 4: "Growth"

The "Growth" phase focuses on a very precise objective: to build promotional gimmicks able to create continuous traffic to your course. These are simple actions that serve to create information content or call to action to convey users towards the purchase of a course online:

a. **Tutorials.** Tutorials are one of the main ways to drastically increase the number of target users with your training product, who visit your course page. Usually, I always recommend the realization of short tutorials (no more than 30 minutes) on topics similar to one of the courses to be promoted. For example, one of my students who created an online course on WordPress has created the first free tutorial to explain how to install WordPress via FTP. This allowed her to disseminate free content able to show users her expertise in this profession and consequently increase the perceived value of her online course. Obviously, within the tutorials, there are references and redirects to the online course.

b. **Blog.** A blogging strategy and tutorials can help you get continuous traffic to your online course. With a blogging strategy, you can talk about your content to your audience and then include references to your online course and teacher profile. The blog is not an exclusive tool for Content Managers, but thanks to the application, exercise, and consistency, as well as the study of basic blogging techniques, you will be able to create your website to push your course!

c. **HelloBar.** HelloBar is a simple promotion tool to integrate on your blog or website. With HelloBar, you can create direct backlinks to your training course, or you can create an opt-in form to use to build and monetize your email database. You can proceed with the free installation of HelloBar by clicking here. If you prefer a complete set of Tools to integrate on your site, you can try *SumoMe.*

d. **Sidebars.** The best way to create a "lead magnet" is to use the sidebars of your blog to insert banners with a direct link to your course page. By inserting the banners, you can keep a marketing expedient directly on your site. By doing so, you don't have to worry about constantly updating your content. Everything is simple, fast, and automated.

Tips On Making The Lesson More Motivating

A video course that sells is an easy-to-navigate video that conveys the information a user has paid for, so organization and planning are indispensable.

To proceed, take software like Xmind or pen and paper and draw the map of your course video as if they were the chapters of a book but deepening the various titles of the content.

Divide the map into modules and sections, where the module is the category, the step, or simply the chapter of the course.

Then take each module and write the important components to develop it; I usually write down the titles of future videos or simply the keywords.

In this way, you will organize your course video better. It will be easy for the user to review certain information.

You should know that a client attaches more value to a course video organized in modules and sections than one created only with long and complex videos.

How to Make Online Courses More Interactive

Making online courses more interactive is key since this increases the motivation of participants. But how?

First, I want to explain to you the meaning of interactive communication. Secondly, there will be some tips to improve the interactivity of online courses.

Mutual communication is a key element with interactive communication. Before the arrival of computers, the internet, and mobile and digital media, there was only one-way communication. There is a sender and the receiver cannot answer. It simply receives the given information. Examples of one-way communication systems are books or television. Interactive communication is an exchange of ideas in which both participants, whether human beings, machines, or art forms, are active and can affect each other. That said, it's time to give you some tips to make your online course more interactive!

Tips to improve the interactivity of your online course:

Promote Group Collaboration

Start group discussions about your online course so your participants can discuss certain questions and work together on projects. You can also add a teacher to the chat. In this way, the teacher can lead the discussion and encourage students to solve problems collectively through a group chat. In this way, the teacher can continue to keep an eye on the participants. Human interaction allows students to learn from each other!

Ask Questions During an Online Lesson

Live online lessons can be a bit boring because students can't ask questions. Involving them is important to increase their involvement. During an online lesson, students can ask the teacher questions. And this is also true in reverse; the teacher can ask the students questions.

Include Quizzes and Tests

Not only receiving an answer from a person but also the media is a way to make a quiz interactive. Including quizzes and tests increase student involvement, and it is a fun way to learn knowledge and allows you to receive immediate feedback from the quiz or test. Thus, the quiz or test will show students their performance.

Chapter 8

How To Publish eBook to Earn Extra Money

Coming Up With Your Book Idea

So, you want to write a book and get it out there for people to read. That's good. Most people have something important to say. However, first, you must answer an important question: do you just want to write this book for yourself and your friends or family to read, or do you want to write a commercially viable book that will sell well?

I find many people just want to write and produce a book for themselves and close friends. If so, then the title you choose and the method you use won't matter much. In fact, you can probably jump to Chapter Three. However, suppose you want to sell a reasonable amount of copies and maybe make some money off this book. In that case, you will need to put some dedicated thought into the project before you even start writing.

I also recommend that you download a pre-formatted Word template through www.KDP.com before you start writing. Writing your book directly into the template will save you a lot of formatting time later since you won't then have to cut and paste into the template from another incorrectly formatted document. To access those templates, first set up your KDP account as described in Chapter Five. Then, on your KDP account homepage, you will a rectangle entitled, "Create a New Title." Within that rectangle, you will see some orange checked circles. The first of these says: "Book Content." At the end of that paragraph, there are words in blue "Get started with Kindle content creation tools." Click on those blue words even if you're not planning to set up a Kindle book right now. On that next page, scroll down until you see the Paperback flow chart. Do not be alarmed by the volume of information

on this page. Just scroll down to the Paperback flow chart under which you should see more blue words. Click on the blue words: "Manuscript templates" buried in what looks like a long sentence of choices. In the alternative, click on the word "Help" at the top right of your account page home screen. Then enter "Manuscript templates" in the search bar at the top middle of the help page. Click on the magnifying glass, and you will see a list of choices. Pick the choice (which should be at the top of the list) in blue, "Paperback Manuscript Templates," by clicking on those words.

Either of those two paths should bring you to a page entitled "Paperback Manuscript Templates" in orange writing. Scroll down until you see numbered steps. You want "Step 1: Choose a template." Within this section, click what, as of this writing, is the second gold Download button marked (Templates with sample content). Doing so will download a zip file to your computer, within which are all the Word templates for any format size KDP will print for you. "Templates with sample content" will show you where your content should go within the template. Blank templates aren't very helpful. For sample content templates, simply cut and paste your content over the sample content.

You will need to decide what size you want your book to be before choosing your template from those you have downloaded in the zip file. You will find that the KDP book setup defaults to books that are 6 X 9 inches. This book, as a paperback, is 5.25 X 8 inches because I wanted it to fit easily in your hand when thumbing through it as a reference. Feel free to go to your own bookshelf and look for a book the size you want for your book. Then, just measure the book. If you're writing a book for a specific genre or market, look at several books in your category.

Decide which size works best for you, but try to pick one of the standard sizes offered within the KDP templates. Then, click on that size from the template list, which should appear when you open your zip file. If the template zip file comes into your download window, just save it to your book file or folder on your computer. If it comes in on the bottom of your screen, click on it, and it should open it to show you all the files within. You should then be able to choose where to save them on your computer. Once you've chosen your book size, click on that template wherever you have saved it on your computer. If you see a yellow band across the top of your screen, click "enable editing" at the right end of that band. Then save the file in the book folder on your computer.

The next question to ask yourself is will your book be a work of fiction or nonfiction? Fiction books are printed on cream-colored paper and nonfiction books on white paper. Most people don't know that but check your own bookshelves. You'll see that's the way it's done in the publishing industry, and it's a great way to make your book look professional.

Choosing your title is very important. You'll want to catch people's attention with your title and intrigue them enough to buy your book if you're writing fiction. If you're writing a nonfiction book, try to think of keywords that people would search for on the internet to find your book. Put as many of those as you can in the title. Don't forget that you can add more of these search words by adding a subtitle. You can do the same with a fiction book. For example, I named a book I published, *The Finger of God: A Young Adult Biblical Adventure.* The subtitle was actually suggested by my proofreader, and it was a good suggestion. Without the subtitle, no one would really know the genre of the book. Subtitles can also help differentiate your book if it turns out there are other books with similar titles - although it's always a good idea to make sure your title is unique before you settle on it.

I named another of my books *A Manager's Guide to Preventing Liability for Sexual Harassment in the Workplace.* Some people told me that the title was too long. However, no matter what combination of search words a manager puts in when looking for a book like mine, my book

generally comes up first in an Amazon search because those search words are in the title. This approach has earned me many sales I would not have made if my book had come up further down the list. So, spend some time thinking about who you want to target as a reader for your book and what words they would type in if they were trying to find a book like yours to read or buy. Then, fashion your title accordingly.

Realize that sometimes longer titles will not fit on the spine of your printed book. However, since most of my sales are made through Amazon where people only see the cover, not the spine, I still prefer to pack the title with search terms that work on Amazon. Suppose you will primarily sell your book through bookstores, where people browse shelves reading titles on spines. In that case, you might want to re-think this approach and make sure that your title will easily fit on the spine of your book.

If you are trying to make money from your nonfiction book, think about possible places to sell the book. Is this a title that museum shops would carry? Can you fashion a speech from the subject matter as a way to promote your book? Would you be comfortable talking about the subject matter of your book during interviews? Does the topic lend itself to smaller articles that could then link readers to the more in-depth discussion you have in your book? Another strategy is to join special interest groups related to your topic, either online (such as a Facebook group) or in person. These groups can be great sources of information and inspiration for your book as well and audience.

If it's fiction, is your story really long and involved enough to fill a whole book? To have a book large enough to put the name and title on the spine, the manuscript should be at least 30,000 words. Often, the first "books" end up being novellas or short stories. Generally, the cutoff for a "novel" versus a "novella" is about 50,000 words, but novels vary widely by genre. Generally, between 75,000 and 85,000 words is a sweet spot for most fiction (the first book of an extremely popular fantasy series, for instance, is 76,944 words), and 100,000 words tend to be on the high end. However, some genres, particularly fantasy, can support higher word counts, especially if your book contains extensive world-building. For example, the longest book in the series above is a whopping 257,045 words. However, bear in mind that most of these door-stopper fantasy books come from established writers. Most publishers won't want to commit to such a huge book from an untried or first-time writer, and, even if you self-publish, it can be difficult to get readers to commit to such a large book from an author without a large public profile.

If it turns out that your book is significantly longer than 100,000 words, there are two ways forward. You can either viciously edit your work to get your word count down (and you should always try to get rid of extraneous content that doesn't move the story forward), or, if you've already "cut out the fat" and are left with only essential story elements, perhaps your one huge book can become two books. For example, I once wrote a book that ended up being 160,000 words. I realized that all of the plot points were necessary, and nothing could really be cut out. However, there was a natural breaking point almost exactly in the middle of the book, which ended up becoming two approximately 80,000-word volumes. Writing a whole novel is quite an endeavor. I find it best to map out the story enough to ensure that you have the necessary chapter count and plans for your character arcs. Finding your writer's voice, developing characters, and creating a viable plot are subjects for another volume. However, these are all important elements of a fiction book that can't be overlooked. Just because you are self-publishing doesn't mean that the book doesn't have to be good.

Many people ask me whether self-publishing precludes the possibility that a mainstream publisher will pick up the book for their production line. Actually, several self-published books recently did so well on their own that publishers then asked the authors if they could make those books part of their offerings. Those traditional publishers then helped those authors catapult to the top of bestseller lists and get movie rights.

One such author released his book online, chapter by chapter, growing his audience with each installment. Then, he released the final chapter only in book form, which his now hooked readers had to purchase to find out the ending. That book later also became a blockbuster movie. So, self-publishing has become a way to get noticed and prove your writing is commercially viable. Self-publishing no longer appears to cut off all hope of finding a home at a large publishing house.

Using a print-on-demand publisher, such as KDP, allows you to buy your books one at a time rather than ending up with a garage full of books like you would if you went to an old-fashioned printer that will usually require a print run of at least several hundred copies. With an on-demand publisher, you can buy as few or as many books as you need, usually at a price per book close to what you would receive if contracting with a full press. You can also make minor changes in between orders so that the next person who buys your book will receive the new version because the book isn't printed until ordered. For self-publishing, I believe on-demand printing is the way to go.

About ISBNs

Next, let's discuss why you need an International Standard Book Number (ISBN). The ISBN is a number recognized internationally to identify your book and its format and publisher. Consider the ISBN like a Social Security Number for your book. There may be many John Smiths, but credit agencies, the government, employers, etc., tell them apart by their social security numbers. Likewise, other books may have a title the same or similar to yours over the years, but no others will have your ISBN. Consequently, libraries, bookstores, Google Books, etc., use your ISBN to identify your specific book, as well as its format. So, if you have a paperback version and a hardcover version of your book for sale, each will need a separate ISBN.

Assigning an ISBN to your book is not a legal requirement to getting a manuscript printed. You can self-publish a book without an ISBN. However, then you will only be able to sell the book privately and directly to customers yourself. Libraries will not be able to find it. Bookstores will not carry it, and Amazon can't list it for sale. So, it's best to have an ISBN assigned to your book, even if you think only family and friends will buy it. Not only does the ISBN make your book more professional, but it also allows it to reach a broader audience. If you don't want to buy ISBNs, you can use a free one offered by KDP, as described in the last chapter.

Before 2007, ISBNs were ten digits long. After January 1, 2007, ISBNs became thirteen digits long. I guess with so many new authors writing books, and they needed more digits to accommodate the production, similar to how our automobile license plate numbers keep adding digits in California.

If you want to be your own publisher, use your own ISBN purchased from Bowker. Bowker is a privately-owned company that distributes and catalogs all the United States ISBNs. Yes, Bowker is a monopoly. Unfortunately, there is nothing we can do about that before you get your book up for sale. So, don't buy an ISBN or a set of ISBNs from anyone or anywhere else. If you do, those ISBNs will have already been assigned to someone else, and that someone will end up being registered as your publisher. To be registered as the publisher of your book, you must buy your book's ISBN directly from Bowker.

To do that, you must first get a Bowker account. So, go to www.myidentifiers.com. You can go to Bowker.com, but that website will just refer you to www.myidentifiers.com eventually and probably confuse you in the meantime.

The homepage on the MyIdentifiers site might seem a bit daunting at first. Bowker would like to sell you many things. All you need here are ISBNs, so ignore the other offers.

Here you will need to decide how many ISBNs to purchase. The more you purchase at once, the less expensive each individual number becomes. Remember, you will need an ISBN for each format you will use for your book. Do you plan on a possible Kindle edition? If so, that edition will need a separate ISBN. I usually recommend that you buy ten numbers the first time. That choice gives you some flexibility and brings the price down from $125 for one number to $29.50 per number for each of the ten numbers you purchase at once.1110f course, one hundred ISBNs for $595 presents the best deal, but what would you do with one hundred ISBNs?

You do not need to purchase ISBNs and their barcodes together. When you enter the ISBN into KDP and eventually use KDP to create the cover for your book, KDP will generate the barcode for you; one more reason to use KDP. Even if you create your own cover, there is a site where you can get your barcode for free. We'll discuss how to create your own cover in Chapter Eight.

Bowker will prompt you to set up your account when you go to buy your ISBN(s). Your email address will become your user name, and you will have to pick a password.

Bowker will prompt you to fill out multiple pages on its site. You only need to enter information in the boxes marked with red. If Bowker asks you to enter your tax identification number and have an EIN other than your SSN, enter the EIN. Also, on the "My Company" tab of your Bowker page, you can enter that in the space provided if you picked an imprint name. Suppose your EIN matches a company other than your imprint name. In that case, your company name goes in the primary space, and you can add additional imprint names below your company name and then use them in your books.

If you see a place for a "SAN" on the Bowker site, just ignore it for now. If you are using KDP as your on-demand publisher, all book orders will actually go through them. So you don't need a SAN, which is a seven-digit number used to identify the publisher's shipping address.

To set up your book for publishing and assign an ISBN to it, click on the "My Identifiers" tab in your Bowker account. You will then see all your identifiers in numeric order. Just start with the first blue button labeled "Assign Title." This action brings you to the page you will need to fill in with information about your book. You do not need to fill in all the information at once. Some you may not even know until you set up your KDP account. For now, you can just assign your title to an ISBN. Then you can use that number to get a Library of Congress Control Number and to set your book on the KDP site.

Once you click on that "Assign Title" button, you will land on a one-page form. In the upper right-hand corner, Bowker has defaulted you to the "Basic" form, which is now only one simple page to fill out. If you wish, you can switch to the "Advanced" form, which has more fields to fill in.

If you stick with the Basic page, you will see your thirteen-digit ISBN at the top left in bold. The first field to fill in calls for your book title. If you have a subtitle, you enter it in the field directly below. If you don't have a subtitle, you can leave this field blank.

At this point, all you really need to fill in on this page is the title of your book. You only need to fill in the boxes on the identifier pages labeled with red asterisks. In fact, you may not yet know the answers to some of the other questions, such as copyright year, page length, publication date, etc., until later. You can go back into your account to add these later. It is better to wait until you have all the information and your book is uploaded to your printer (KDP or equivalent) before you fill out those extra boxes.

Once you have entered the title of your book, you can then enter the book description in the space provided. I usually cut and paste in the description I use for the back cover of my book. To the right, there is a place to upload a jpg file of the front cover of your book, if you wish.

The next two fields are mandatory. You must pick a book medium from the drop-down menu. Pick either Print, E-Book, or Audio, depending on the format you will assign to this ISBN.

Next to "Medium," you'll find another mandatory field, "Format." Pick the format that best corresponds to your medium. For example, will this ISBN be assigned to the paperback or hardcover version if it's a print book? You will probably choose "Paperback" in the "Format" drop-down menu. If you plan to publish first in a Kindle or other e-book format, choose "E-Book" in the "Medium" drop-down menu. Then you will only have one choice in the drop-down menu for "Format," but you will gain a field called "e-Book File Type." Here, a safe choice will be E-"Pub."

Next, you must pick at least one genre to describe your book. Pick the one from this drop-down menu that best describes the genre for your book. You can add a second genre in the field to the right if you wish.

Next, you will list "Contributors." You can add your name as the author. It will ask your first and last name and allow you to list a suffix, such as Ph.D. if you wish, and it truly applies. You'll see a box in which you can enter up to a 350-word biography of the contributor just below where you entered the name. Underneath that box, you will check that contributor's function, such as author. Suppose your book has more than one contributor. In that case, you can list others by clicking on the blue "Add Another Author/Contributor" button just below.

Next, you'll come to Sales and Pricing. You'll see that Bowker now auto-fills the publisher field based on the information in your Bowker account. Next, the publication date must be filled in before you can submit the form to Bowker. You will not fill in this field until you have actually published the book and have that date. Once your book is for sale on Amazon, you can find the date by viewing your book on Amazon and scrolling down to "Book Details." Under that heading on the page that displays your book for sale on Amazon, you will see the publication date and other useful information such as the size, page count, and weight of your book. You will also see the publisher of your book listed there.

Don't worry that you can't fill in the publication date yet. You can save the form information and come back to add that final bit of information after you've published the book. However, make a note to come back and actually submit the form once your book has been published. It's easy to forget this step in the euphoria of seeing your book in print.

Under "Publication Date," you'll see another mandatory field: "Target Audience." Here, select the best answer from the drop-down menu. Unless you have a specific target audience listed there, you can choose "Trade." Next, for "Title Status," you can choose "Forthcoming" while the book is in production, but make sure to change this field to "Active Record" once you've published the book.

The last required field is to enter a dollar amount for the sales price of your book. Enter the retail price you have chosen for your book. To deter

mine a price, I usually check to see what others are charging for similar books on Amazon.

If you wish to add more information, you can switch to the "Advanced" version of the form by clicking on that button at the top right-hand corner of the page. The Advanced form allows you to enter even more information about your book, such as the Library of Congress Control Number, if you've got one (See the next Chapter), the copyright year, and more information about the size and formatting of your book. Suppose your imprint name is different from the publisher name that auto generates in your fields. In that case, you can enter an imprint name on the fourth page under the Sales and Pricing section of the advanced form.

Bowker has helpful explanations of their fields accessible by clicking on the blue "?" next to most fields. These will help you to understand what they expect from you concerning each field. Just follow their instructions if you need extra help.

If you need to leave the Bowker site before you have filled in every field, just make sure that you are comfortable assigning that title to that ISBN and hit the green "Submit" button at the bottom of the Sales and Pricing page before you sign off. You can go back in and change most of the other information later if you wish.

To re-access your book information later, just go to the www.myidentifiers.com site again. Click on "Log In" in the upper right-hand corner of your screen. Then enter your email and password to gain access to the site. Once you are in Bowker, put your cursor over the My Account tab, which lies under the Bowker header but near the top of the page. You'll see a drop-down menu appear, then click on "Manage ISBNs" under "My Identifiers," which is the first on the list.

If you don't fill in all the fields marked with a red the first time you assign the ISBN, don't forget to go back and fill in those fields as soon as you know the publication date of your book.

Setting Up Your KDP Account

Given all the choices of on-demand publishers, I preferred CreateSpace, which Amazon has recently discontinued in favor of Kindle Direct Publishing or KDP. However, KDP still offers a free set-up of your book for printing. Also, as an arm of Amazon, it is designed to work seamlessly with Amazon.

You will find that there are "vanity presses" which offer to publish your book for a fee. I often find these fees exorbitant. Please avoid vanity presses. If you follow the guidance in this book, you will find that there is little a vanity press can do for you that you cannot do for yourself for free. Alternatively, you can hire experts for discrete tasks, such as editing, for smaller fees than may be charged by a vanity press. Vanity presses may claim that they will help you with marketing. In reality, you will retain most of the responsibility for marketing whether you use a vanity press, a traditional publisher, or an on-demand publisher/printer.

An account with KDP is free to set up. There are no upfront charges for your book. You can upload edited or updated versions also free of charge. Therefore, you can write and upload your book to KDP for free, and because KDP is a subsidiary of Amazon, your time will automatically be placed for sale on Amazon once you are happy with the proof version and hit the "publish" button. So, getting your book to show on the proper Amazon lists is easier, and you can easily convert your book to a Kindle edition if you work through KDP.

In addition, KDP offers expanded distribution channels for free, meaning that bookstores can find and order your book for their shelves, as can libraries. Purchasers can find your book for sale on other online channels as well, such as barnesandnoble.com or bookshop.org. So, I see no downside to using KDP as your on-demand printer, although some features are not as easy to use as I found them on the CreateSpace platform that KDP replaced. Please note that, as discussed in the previous chapter, I personally prefer to keep KDP as my printer, not my publisher.

Setting up your KDP account literally takes just a couple of minutes and is even easier if you have an existing Amazon account. To set up your KDP account, go to www.KDP.com. When the page loads, you will see that you can sign in using those credentials if you already have an Amazon account. If you do not have an existing Amazon account, simply click the "sign up" button under the words "Don't have an Amazon account" located toward the upper right of your screen. If you do this, it appears to take you to a screen where it again says, "Sign in with your Amazon login." If you see this and don't have an Amazon login, don't panic. Just click on the grey "Create your KDP account" button at the bottom of that box. You will be prompted to enter your name, email, a password and to confirm your password. Then, click on the gold "Create your KDP account" button. If you already have an Amazon account, just click the gold "sign-in" button that you will see under the words, "Sign in with your Amazon account." Then, follow the prompts.

Once you do this more than once, KDP will remember you, and all you will need to do is click that Sign in button and be directed to the KDP home page. However, the first time you enter KDP, either by setting up a new account or using your existing Amazon account log-in credentials, you will be prompted to agree to the KDP terms. Feel free to read all this fine print before agreeing, but you can't negotiate the terms. If you want a KDP account, you will need to click on the gold "Agree" button at the bottom right of the page.

Once you agree, you will be redirected to the home screen of the KDP website. Suppose this is your first time on the KDP website, whether or not you are an Amazon user. In that case, you will likely see a rectangle at the top of the page saying that your account information is incomplete. If so, then click on "update now." You will then be prompted to fill in some information. First, choose the country in which you work. Click on the little arrows next to the empty "Country" box to see a list of choices. The United States is at the top of the list. The rest of the country's choices are in alphabetical order.

Once you click on a country, a new page will open with boxes to fill in your publisher information. Fill in your full name, address, and phone number. Under that is a box entitled "Getting Paid." Use those little arrows again to say where your bank is located. To get your royalties directly deposited into your bank account, you will need to fill out all the information required in this box, which will expand after entering the country in which you bank.

If you do not have a bank account or do not wish to receive your royalties by direct deposit, you can try clicking on "I don't have a bank." When I did that, it said that my jurisdiction required me to have a bank for payments. This response indicates to me that Amazon and KDP don't want to issue checks anymore and are trying to force us all to accept direct deposit payments. Hopefully, you have a bank account that you don't mind entering here so that you can continue to set up your KDP account. I actually enjoy getting my royalties directly deposited in my business bank account. It saves me a trip to the bank to make the deposit.

Under the Getting Paid box, you will find the Tax Information Box. Click on the gold "Complete Tax Information" box and answer the required questions. Your royalties are income, reportable to the IRS. So, you will not sell books through Amazon using KDP as your printer if you do not fill in the required tax information. Suppose you took the trouble to get a separate EIN for your publishing business. In that case, you can enter the EIN here instead of your Social Security Number.

Once you have set up your account, you'll be directed to the KDP homepage. If you exit KDP and want to come back into your account, just go to www.KDP.com and click the gold "Sign in" button. You'll see that the site will most likely remember you if you access it from the same computer you used to initially sign up for your account. If the site doesn't remember you, just enter your email address and password as prompted and click the gold

"Sign in" button. You will find yourself at the KDP homepage again.

Prepping Your Book For Publishing

Once you have set up your account on KDP, you can start loading your book into the KDP system for publication. On the KDP account homepage, you'll see that the third box down is labeled "Create a New Title." Click on the box within that rectangle entitled "+ Paperback."

KDP may ask you to sign in again. If so, just do it. Then, you will come to a page with three large tabs at the top. The left tab, "Paperback Details," will say "in progress . . ." That is the tab that will be open for you to work in. There are three tabs with information in boxes that you must complete before your book is ready to publish on KDP, or rather before KDP is able to print copies for you on-demand and list your book for sale on Amazon. For ease in following these steps, I add a label that corresponds to each box you must fill in with corresponding instructions under each label. For example, the first box you will come to is labeled "Language."

The Language Box:

In the language box, choose a language for your book. This box usually defaults to English for you.

The Book Title Box:

Enter your book title in the space provided. As discussed in a previous chapter, if you have a subtitle, enter it in the space provided. A subtitle is not required, but it does allow you to add more internet search words in your titles, allowing more people to find your book when they look for your topic.

Once you have entered a title, you should be able to come back to the setup process even if you exit all the way out of KDP. If you need to exit KDP and find yourself back on the homepage when you log back into your account, just scroll down until you see the words, "Your Books." You should see the title of your book listed there. To the right of that title, you should find a gold button entitled: "Continue setup." That should take you back to where you left off in the setup process.

The Series Box:

Most first-time authors will not be writing a series or a second edition. If that's true for you, just leave those boxes blank. If your book is one in a series or you plan for it to be the first in a series, enter the number which corresponds to where your book falls in the series into this box.

The Edition Number Box:

If this is not the first edition, then fill in this box accordingly. Most likely, this is your first edition of the book. If so, then you can leave this box blank. You will see that it is marked "optional."

The Author Box:

Enter yourself as the primary author, assuming that you are the author. If you are not the author and will be publishing the book for someone else, then enter the primary or first listed author of the book here. You must list at least one primary author or contributor. Some books don't actually have a primary author. Instead, they have an editor that complied with the works of several authors. In that case, enter the editor as the primary contributor in the "Author" box.

The Contributors Box:

If you have other contributors, such as an illustrator or co-author, you can enter their names and functions in this box. Use the drop-down menu from this box which defaults to "Author," to identify which contributor you will enter. If you have more than one contributor, you can use the "Add Another" button to add the additional contributors. Chances are that you will not have any other contributors, so you can just skip this step altogether. Just to clarify, if you hired an editor to read your book, comment on it, or proofread it, that type of editor is not a contributor and should not be listed here.

The Description Box:

Next, enter your book description. Here you have the option to enter up to a 4000-character description of your book. Take advantage of this opportunity to put keyword search terms into your book description. I often use the same description I place on the back cover of my book. However, KDP allows you more characters for your book description than for the brief description that will fit your back cover. So, if you wish to expand the description, go ahead and write more in this description box. Whatever you enter in this box will appear on Amazon as the book description. So, if you are unsure what to write in that box, go to Amazon and look up books similar to yours to see how they are described.

The Publishing Rights Box:

Under the book description box, you will find the "Publishing Rights" section or box. Here, pick "I own the copyright, and I hold necessary publishing rights." If you wrote the book, you own the copyright whether or not you've registered the copyright with the Library of Congress.

The Key Words Box:

KDP allows you to enter keywords to describe your book. To choose the keywords, think about how people would search for your book online if they didn't know that your actual book existed. What is it about? What type of people will benefit from finding it? What type of people will want to buy it? You can enter up to seven words or short phrases. If you enter short phrases, please put quotation marks around the phrase.

The Categories Box:

Here you can choose two browse categories for your book. Some people search Amazon by category, so this is one chance to make sure that your book gets in the category that best defines it. These categories really search paths that narrow the search as they go. So, if you pick non-fiction as the starting point, you will see that you will have lots of sub-categories under non-fiction from which to choose. Keep choosing increasingly narrow sub-categories so long as you are clicking on a category that has a plus sign in front of it. The narrower you can make your category, the better because you will have less competition.

You will also see a box that you can check to indicate that your book is printed in large print. Only check that box if your book's regular prose is printed in fonts of sixteen points or higher, not counting headers or chapter titles. If your book's fonts are twenty-four points or higher, then most consider the book as a "giant print" book.

The Adult Content Box:

In this last box, under the first tab, you need to state whether or not your book has adult content. Essentially, they are talking here about X-rated content - the kind you would find in an "Adult Bookstore." So, you don't have to have written a children's book to click "no" in this section.

Once you have completed all the information under the "Paperback Details" tab, you can click the gold "Save and continue" button at the bottom of the page. This action will take you to the next tab entitled "Paperback Content." I will also label each section from this tab so you can more easily follow the instructions.

The Print ISBN Box:

The first question under this tab is whether you want to use a free ISBN supplied by KDP or provide one of your own that you have purchased through Bowker on the Myidentifiers.com site. We discussed the pros and cons of that choice earlier in the book, in the first chapter, and how to purchase your own ISBNs in Chapter Three. If you have decided to truly publish your book for free and that you don't mind KDP being forever listed as the publisher of your book, then go ahead and click the bubble next to

"Get a free KDP ISBN." Instead, if you decide to click on "Use my own ISBN," a box will appear in which you will enter one of your own ISBNs. Make sure you write down which ISBN you used if you own more than one so that you correctly assign that number to the right title in both KDP and in your MyIdentifiers.com account. Below the ISBN choices, you will need to enter your imprint name if you've entered your own ISBN. Choosing your imprint name was also discussed previously in Chapter Three, which covered getting your ISBNs. If you've used a free KDP ISBN, your book will be printed under the KDP imprint.

The Publication Date Box:

After you fill in the ISBN box, you will come to the publication date box. At this point, you may not know your publication date. Some authors schedule the release of their book in advance. If you plan to do that, you can list the publication date there and make sure to submit the book on that date. Otherwise, you can leave this box the way KDP has already filled it in for you. KDP has already set this box to fill automatically on the date that you make your title "live" on Amazon - in other words, the day that you ultimately hit the "publish" button at the end of the process.

The Print Options Box:

Beneath the Publication Date box, you will find the "Print Options" box. Here, choose a black and white interior with cream paper if your book is fiction or a black and white interior with white paper if your book is nonfiction. The paper color is not a matter of personal preference. Traditionally, non-fiction books are printed on white paper and fiction on cream paper. Check the traditionally published books on your own bookshelf if you don't believe me. If you are publishing a children's picture book or a book that will include color pictures, then choose the print option. However, please understand that a color edition of your book will cost much more to produce than a black and white version. So, if your pictures will work in black and white, stick with black and white printing.

Next, you will need to pick a trim size for your book; Trim size refers to the outside dimensions of your book. KDP lists the width first, the dimensions across the front cover, and then the height; the top cover is measured from top to bottom and does so in inches. You will see that KDP defaults to the most popular trim size, 6 x 9. If you prefer a different size, like I did for this book, then just click on the box to the right of the one where it shows the dimensions entitled, "Select a different size," and choose your preferred trim size. Feel free to take a ruler to any books on your shelf that seem like the right size to you.

There are quite a few trim size choices offered by KDP. Think about how long your book will be. Should you choose a smaller size so that you have a longer page count? A note to the wise here: KDP requires at least a hundred-page book before printing the title and author name on the book's spine. Do you want the freedom to make the type larger? If so, pick a larger size book. What size did you picture in your mind when you first conceived of your project? What do traditional publishers use for books like yours?

I usually pick 6 X 9 for nonfiction textbooks unless the book is more of a manual. Then, I choose a smaller size to feel like it fits in the palm of a hand. I sometimes also go smaller for fiction, using 5 $^{1}/4$ X 8, the trade paperback size. It's really up to you. However, pick one of the trim sizes offered by KDP. Such a choice will make the whole process much easier.

Unless your book will include pictures that will cover the whole page without any white space around them, click the no bleed option under "Bleed settings." The bleed setting is only available for print-ready files, such as pdfs, according to KDP. Then, you can decide if you want a glossy or matte finish on your cover. I prefer glossy, but this is totally a matter of personal preference.

The Manuscript Box:

Once you've filled in all the boxes in the Print Options section, you'll come to the Manuscript section. Here, in the fine print, you'll see that you can download a free Word template for the trim size you chose above. Please do so. Just click on the blue words, "download a free KDP template." Doing so will take you to a page entitled "Paperback Manuscript Templates." Here, you will first see a video link about how to use the templates. Watch that video if you wish, or skip down to where you see "Step 1: Choose a template." Click the gold button marked "Download" next to "(Templates with Sample Content.)." You will want to use a template with sample content to make the process easier.

When you click that gold button, you will download the templates for all the trim sizes as a zip file. Below those buttons, KPD lists all the trim sizes it supports. Once you choose your trim size from that list, you can open the zip file, choose the template from the list that opens on your computer, and then click the one you want to work on.

If you see a gold band at the top which says, "protected view," click on the right-hand end of that line to "enable editing." Then, you should be good to go.

If you have not already written your book, you can write it right into the template. The template will come pre-formatted for your trim size with a title page, table of contents, chapters, etc. Basically, all the components of your book are laid out for you in the template. You literally just need to fill in each section. Make sure to replace the sample text with your own in each section of the template. For more instructions on how to do that, watch the video provided by KDP. I will also walk you through how to use each section of the template in the next chapter.

Suppose you've already written your manuscript by the time you read this book. In that case, you can just copy from your manuscript and paste it into the template, but make sure to paste it into the template without prior formatting to retain the template formatting for your book. To do so, copy the text in whatever way you normally would. To paste it into the template, place the cursor where you want the copied text to start. Then, go up toward the top left of your screen to the clipboard-looking icon that says, "Paste" under it. Click on the little down arrow under the word, "Paste." Then click on the choice, "Paste Special." In the box that pops up, choose "unformatted text" and click the "Okay" button at the bottom of the pop-up window. Suppose you don't paste in this way. In that case, the regular paste command will paste the text you highlighted and all the formatting related to that text. Importing another formatting into an already formatted template can cause a multitude of problems. Please save yourself a lot of grief and use the special paste command with the "unformatted text" option.

One word of caution here relating to the actual text of your book: make sure to use a font that has a big "0" next to it on your font list. This type of font works best when you go to upload your book to KDP. Most of the KDP templates default to Garamond 11-point font. If you want to make sure that you have fewer issues when you upload your interior file to KDP, write your manuscript in that font and use that font for your book. It will likely save some formatting work later on.

To find out how to fill in the rest of the template besides the book's text, please go to the next chapter. Since instructions for formatting your book, designing your cover, and previewing your KDP book file each require chapters of their own, the rest of the boxes under the KDP set up tabs that appear after those subjects will be discussed in Chapter Ten entitled, Rights and Pricing.

Formatting Your Book before Uploading

The process of filling in the KDP template can be fairly straightforward but requires care. Enter your book title on the "Title" page and your name as the author below the title. If you have a subtitle, enter it directly beneath the title in a font smaller than your title but larger than your author name. Make sure that you leave a line or two of space between the title and the subtitle. You can use this book as an example.

The template is laid out so that the left-hand page you see on your screen will be a right-hand page in the book. So, the title page, which shows on the left in the template view, will end up on a right-facing page in the book. This can be disorienting and confusing when working in the template, so periodically check to make sure your "facing" pages are actually where you want them.

The page which appears next in the template will be on the backside of the title page. It is officially called the colophon. You've probably never heard of a colophon. Don't worry, and I hadn't either. That page is also known as the copyright page or the edition notice. On the colophon, you place all that important information with nothing to do with your story or topic, such as the copyright notice, the ISBN number, your imprint, edition number, disclaimers, etc. Please look at the colophon in

this book for an example of how to lay it out. The KDP template gives you placeholders for the bare minimum you need to put on this page. Suppose you add a Library of Congress Control Number, a disclaimer, and your imprint. In that case, your book will look a lot more professional. Note that many of the KDP templates include text toward the bottom of the page and will not let you add content above it. Don't be alarmed. Suppose you start typing in and adding your content to where it already is toward the bottom of the page. In that case, it will move the text up to accommodate additional content.

It is always a good idea to add a disclaimer, especially if anything in your book could be considered professional advice. You are writing a nonfiction book. If you are writing a work of fiction, this is where you make that statement that all characters are fictitious, and even if they seem like people you know, they aren't real or something to that effect. Whether the book is fiction or nonfiction, I also say that no one is allowed to copy pieces of the book for any reason without prior permission. I give my email as a way for people to ask for that permission. If someone wants to write a great review of my book and quote an extremely well-written excerpt from my writing, I want them to be able to ask me for permission to do so. Note that you should make sure the disclaimer is actually true. For example, suppose you're writing a work of historical fiction that includes real historical figures. In that case, you can't say that they are entirely fictional. However, you can say that your presentation of the characters is fictional and that the plot points are fictional if they are. Look at other similar books' disclaimers for guidance on how to write one for your particular genre.

Before CreateSpace was disbanded in favor of KDP, we could say "Printed in the U.S.A." at the bottom of the colophon. While most KDP books sold in the U.S. are printed in the U.S., books for other markets (such as the EU) might be printed elsewhere. The last page of your book will include a statement from KDP about the book's printing location, so there's no need to include that information in your colophon, especially since you can't be 100% sure where your book will be printed.

Next in the template, you will see a page with a space holder for a dedication. You can dedicate your book to whoever you want and for whatever reason. You don't have to put in dedication, but most authors do.

KDP next has a place for you to list your acknowledgments. You will notice a blank page between the dedication page and the acknowledgment page. This blank page allows the special pages to appear on the right-hand side, or facing, page of the book. Most readers expect to see those pages on the right-hand side. Suppose you have other content to put toward the front of your book, such as a map for a fantasy book. In that case, you can use the space reserved for acknowledgments for things other than acknowledgments and include the acknowledgments at the back of the book, as many professionally published books do.

Similarly, I've noticed that KDP does not give a page for a preface, introduction, or foreword. If you wish to include one of these, you can put it on the page reserved for the acknowledgments page at the front of the book. Just make sure to change the title of the page. You can then add an acknowledgments page at the end of your book to have both an introduction and an acknowledgments page. Alternatively, for a longer-form front matter like a prologue, you can use the first chapter, but be careful about page numbering (see below).

Don't try to add a new page before the first chapter of the book in the template. You will notice that the pages before Chapter One are numbered with small Roman numerals. Suppose you add any extra pages before the placeholder first chapter, which is set up in the template, to start on Arabic numeral one. In that case, all the paging will be thrown off for the whole book. So, if you want to add extra sections, like an acknowledgment, add them to the back of the book, not the front, which will not throw off the page numbering.

If you choose to have an acknowledgment page, list anyone that helped you with the book on this page. Usually, you will have friends, family, a proofreader, possibly a content editor, maybe a muse that impacted your final book content. People love to see their names in print, so a generous list here may actually help you to sell your book later. However, you do not have to include either a preface or an acknowledgement page in your book.

The template also has a contents page layout for you, with up to ten chapters listed. However, the template will not automatically fill in the page numbers for each chapter in the table of contents. You will need to add or write your book into the rest of the template and then go back to manually add the page numbers to the table of contents once you have finished formatting the book. If you wish to add chapters, please realize that the table the contents' contents up in columns. So, the chapter number is one column, the chapter title is another column, and the page on which each chapter starts is yet another column. To add chapters, you need to add to each of those columns independently. For example, to add a chapter eleven, first put the cursor after the number 10 in the table of contents and then hit "enter" or "return" on your computer. That will take you to the next space down in which you can add the number eleven for the eleventh chapter. Then put the cursor back at the end of the chapter name for chapter ten and hit "enter." Then, you will be in the right place to add the chapter title for chapter eleven. Repeat the same process to get to the right place to add the starting page for chapter eleven. Many fiction books, particularly ones with a large number of short chapters, do not have a table of contents, so you don't have to include one if it doesn't work for your book.

I usually add my previously written book into the template chapter by chapter to help preserve the formatting in the KDP template. That is also the process recommended by the KDP video. If you are cutting from one document and pasting it into the template, remember to use the "Paste Special" command under the "Paste" dropdown menu on the left, under the Home tab at the top of your Word screen as described in KDP templates have ten chapters. If your book has more than that, as it likely will, simply copy the entire block of placeholder chapters in the template and paste as many times as necessary to give yourself the requisite number of chapters. Just remember to change the chapter numbers when you add in the new chapters.

The template is formatted in sections, so be careful not to change the section settings, or you may find it hard to get your book looking like it should. Don't despair if that happens. You can always use the `undo' arrow at the top of the page to get you back to where things were looking good as long as you don't save the file or sign off before using the undo arrow. The undo arrow is found at the top left of your Word screen. It's a rounded blue arrow pointing to the left and undoes, in order, the last changes you have made to the document, one change for each time you click the arrow.

However, these section breaks (which are very important for things like chapter header formatting) are not visible in the normal editing mode. To make sure you don't accidentally get rid of any section breaks and help you ensure consistent formatting, we recommend you work in preview mode. To do this, hit "file," then "Options," then "display," then, under the "always show formatting marks" section, click the check box next to "show all formatting marks." Now, a lot of extraneous formatting marks will be visible, including little paragraph symbols for each new paragraph and little dots between the words. This makes things harder for you to read, but it also enables you to visually see things like how many spaces you have between words and the locations of section breaks. This will be enormously helpful.

If you were taught to put two spaces between a period and the start of the next sentence, as I was, be aware, that is no longer the rule. Since text wraps digitally now, you should only place one space between the end of one sentence and the beginning of the next. Otherwise, you may find a line indented by one space where you don't intend that to happen.

You will notice that the KDP template puts the book's title as a header at the top of the left-hand page of your screen and the author name at the top of the right-hand side of your screen. You will need to fill these in with the right title and author name as well. Again, check the books on your shelves at home, and you will see this same format in traditionally published books. Just double-click on the book title placeholder text to open the header. Then, highlight the placeholder text and replace it with the title of your book. Do not add the subtitle here. Only use the title and make sure that it types in as all caps like the sample text. Next, do the same for the author name placeholder text, but replace that text with your name or the name of the book's author. This is possible because the template is set to have different headers for even-odd pages. If this formatting ever disappears, double-click the header, and the header "design" tab will appear in the top ribbon, including checkboxes for your settings for headers on even and odd pages. Also, note that these headers typically do not appear on the first pages of chapters. If you lose this setting or want to change it for each chapter, a separate section, double click the header to open the design tab and then check the box next to "different first page." This means that your chapter's first page can have different header formatting than the rest of the chapter.

Also, note that the KDP template starts the text of each chapter on a line that is not indented. That is also the way books are traditionally typeset. Sometimes, the first letter of the first word of a chapter is larger or more ornate than the rest. You can do that if you wish, but it's not necessary. However, especially in a fiction book, these fancy letters, called "raised" or "dropped" initials depending on if the fancy letter towers over your first line or is embedded within the text, can really make your book stand out and look professional. However, to achieve this effect, you can't simply make the first letter a different font or size if you do that, it will mess up the line spacing. The fancy letter must be manually inserted as WordArt. WordArt essentially creates an image.

There will be a button under "Insert" in the menu at the top of your Word screen. On my Word version, I found that button toward the right-hand side of the insert ribbon in the section marked "Text.

Within that "Text" section, an "A" is slightly tilted to the right. That is the icon for WordArt. Click on that icon. Then, type the letter you want into the box that appears. Next, select the font and size that you want and any other special effects like bolding or shadow. Once you've done this, you can drag the letter into place.

Next to the WordArt is a little box that looks like an arc over blue lines. If you hover over it, it will say "layout options." Under this, select "square" or "tight" within the "text wrapping" section. This will enable you to embed your letter within the text without messing up the line spacing.

You will have to manually do this for each chapter. So, rather than starting over each time, it's easier to copy the WordArt and paste it into the right place in each chapter. This will preserve all of your formatting changes, and all you need to do is change the letter by clicking inside the WordArt box and typing in the letter you want. WordArt works like a text box if you are familiar with that.

Note that since your initial letter is now a WordArt, the rest of the first word will not have the first letter. This arrangement will cause Word to think the word is misspelled. To remove the red underlining that appears to indicate a mis-spelling, right-click, and hit "ignore once.

When selecting a font for your book, it's important to bear in mind both professionalism and legibility. Sans-serif fonts (fonts without curvy flourishes on the letters, like Helvetica or the Word default, Calibri) look clean and modern, but studies have shown that they are actually less legible for long books. So, if you are writing a text-heavy book, especially a fiction book, you want a serif font. While Times New Roman is the most famous serif font, everyone can recognize it. If you choose it for your book, it's a bit of a giveaway that your book was self-published. Some popular sans-serif fonts that are a bit more professional, but still included in most common word processors, are Garamond, Palatino, and Baskerville. So, consider one of those.

For your chapter headings, titles, and raised initials, you can afford to be a bit more creative. Choose a font that matches the genre of your book. For example, an action thriller usually won't have a frilly cursive script like you might find in a romance novel. For an extra-custom look, find a font that's not amongst the pre-loaded fonts in your word processor, but make sure that the font is either open-source and available for royalty-free commercial use or purchase an unlimited commercial license if necessary. Generally, fonts already included in the word processors are open-source and free for unlimited commercial use, but they are also commonly used and recognizable, and maybe a giveaway that your book is self-published.

As for font size, generally, fiction books' fonts range from 11 points to 12 points for primary text. Line spacing is commonly 1 point or 1.15 point. To help you pick a combination of line spacing and font size that you like, print several different combinations out, using the same selection of text from your book, and compare them.

If your book includes photographs or illustrations, you can paste them directly into the KDP template wherever you want them. Bear in mind that more images will make your book more expensive to print and thus will force you to set a higher retail price. However, illustrations can really add to a book's appeal. Unless you want to end up with a very expensive book, these illustrations should be black and white/ grayscale. To help you better visualize how the book will look with black and white pictures, it's best to insert them in grayscale to begin with. You can also add to your book's design by including symbols, emblems, or illustrations as part of your chapter headers.

At the end of the book, you will find a place for the Author's Biography. Here you can write all you want about yourself. I find it wise, for future marketing, to put in actual links to my other books and a link to the place on Amazon where readers can review my book.

Granted, you will have to go back and add this link after your book is published and is up on Amazon. When that happens, just go to www.Amazon.com and type the name of your book into the search bar. When you find your book on Amazon, scroll down to the place where it says, "Customer Reviews." Click on the box marked "Write a Review." When your computer loads that page, go to the URL line at the top, and highlight the whole URL. Then, copy the URL. Go back to your book file and paste the URL right into the file after the words, "Please review this book on Amazon at: " Although people will not be able to access the review page directly from your paperback edition, they will at least have the link to type in if they wish to do so. If you later put your book into an e-book format, the link will be there and, assuming Word automatically creates a hyperlink, already live. If the URL is blue and underlined, Word has probably already created a hyperlink for you. To check, just hold down the control button on your keypad and simultaneously click on the URL. If your computer takes you to the review page, then your link is live. If it doesn't, highlight the URL and take your cursor to the "Insert" tab at the top of your Word screen. In that tab, click on "Hyper-link." In the pop-up box, enter the entire URL in the "address" box. I usually just paste it in that box. Then, click "OK" at the bottom of the pop-up window. You will have just created a hyperlink. A word to the wise: do not upload your book as a Kindle edition until you have filled in all the URLs and made sure they go to the review page when clicked on.

When entering content into the KDP template, it's pretty easy to make mistakes or accidentally change some aspect of the pre-loaded formatting. Here are some things to check over before you're ready to upload your manuscript. Are all of your paragraphs indented the same amount? Are all of your chapter headings formatted the same way with the same number of lines between the top of the page and the chapter title, for instance? Do your page numbers show up the way you want them, and do they properly run in their numbered sequence? It's fairly common for section break issues to result in the page numbering starting over somewhere deep into your book if you're not paying attention. Do your headers show the way you want, alternating between title and author without any header on the first page of chapters? Is your font the size and spacing that you want on all the pages of your book? Are the page numbers for your front matter correct? Do you want page numbers for your afterword, bio, or acknowledgments (some books have them, some don't - it appears to be a matter of personal preference). It's easy to tear your hair out when checking through all of these seemingly minor issues, but it's definitely worth it if you want a really professional finished product.

When you have your manuscript entered into the KDP template, and you think you are satisfied with it, consider converting that Word file to a pdf file. You will note that

KDP prefers you to convert your manuscript to a pdf file for a print paperback (this is different for Kindle - more on that later). You can probably upload your file just fine as a Word file. However, if you have many pictures in your book or a really large file for some other reason, you may find it best to first save your file as a pdf and then upload the pdf file to KDP.

You should save your Word file as a pdf if you have one of the newer editions of Word. Simply go to "File" at the top left-hand corner of your Word screen. Click on the File tab in the blue line at the top of your screen. You will see a blue column running down the left of that page on the page that opens. Go down to where it says, "save as," and click on those words. A new window will open. When you choose where to save the document, another window will open to choose the file name and the file type. The file type defaults to "Word." However, if you click on the down arrow at the right of that line, it will open a list of other file types. Scroll down this list until you see "pdf," and click on "pdf." Then click "Save" at the bottom of that window.

Once you have saved your manuscript file in the format of your choice, go back to your KDP account and the "Paperback Content" tab for your book. Scroll down to the box entitled: "Manuscript" and click the gold "Upload Paperback Manuscript" button. That action should open a window on your computer from which you can find your manuscript file. Double click on your manuscript file and wait about a minute. Your computer should take you back to the KDP account, where, eventually, you will see a message that your manuscript file has been uploaded. When you see that message, you can go to the next chapter, describing how to create your cover.

Preparing The Cover

You can spend lots of money on your coverage and have a professional artist create a picture. A graphic artist then incorporates that image into a professionally designed cover. You can just use one of the cover templates and pictures provided by KDP for free in the alternative. You can also use a KDP template with one of your own pictures. Yet another alternative is to disregard the KDP template and create your own cover using another software platform. We'll discuss all of these options here.

Suppose you hire a professional book designer and/or cover artist. In that case, you will likely end up with a very high-quality, premium product. It will also save you a lot of work. Depending on your artistic skill level, a professional artist may be able to visualize your cover better than you could. However, this can be very expensive, especially for book projects without high anticipated sales, like memoirs. In addition, if you hire a professional, you lose some control over what your book will end up looking like. No matter how good a graphic designer may be, he or she can't get into my head to see the image I've envisioned. I'd rather pay nothing and get close to my vision by myself than pay a lot and still not have a cover that replicates my vision. Creating your own cover is possible to do for free using commonly accessible tools.

Of course, you may feel you have no artistic sense, and money spent on a graphic artist is money well spent. If that sounds like you, search the internet or ask other author friends for suggestions of good cover designers. Watch that you get good recommendations. I've heard of many people who have overpaid for this service and have not been happy with the results.

One way to create a cover for free, and the easiest, is to use the built-in KDP cover creator. If you decide to go this route, log in to your KDP account. Click on "continue set up" under your book title on your KDP account home page. Then click on the paperback content page tab, and scroll down to the "Book Cover" box. Fill in the bubble next to "Use cover creator to make your book cover." Then, click the gold "Launch Cover Creator" button.

You will then see a Window called "How to Use Cover Creator." To start using the tool, click on the gold "Continue" button at the bottom of the page. On the next page, click the left-hand box called "From Image Gallery," unless you plan to use one of your own photos on the cover. If you want to use your own picture, click on the middlebox entitled "From My Computer."

If you are using a picture other than one provided by KDP, please make sure that you have the right to use the image. Most pictures found on the internet are not considered in the public domain, so either use one you've taken yourself or that KDP has provided unless you have written permission to use someone else's picture. Suppose you do use someone else's picture. In that case, you should give attribution for the cover image on the copyright page or colophon behind the title page of your book.

If you decide to use a picture provided by KDP, click on the "From Image Gallery" box. You will then be presented with categories of pictures down the left side of the window and picture groups on the right. Click on whatever options intrigue you until you find a picture that captures your vision for the book cover.

You can look through all the images by category to find the picture you want, or you may enter search terms in the search box at the top right of the page. I used "writing," "books," or "keyboards" when searching for a cover picture for the first two editions of this book. You can try as many pictures as you want on your cover. Work on this until you pick one that makes you happy.

Once you have picked a picture, KDP takes you to the cover layout choices for your book. Don't worry; you can always go back and pick a different picture or different layout until you are happy with the cover. For a first cut, just pick the cover template that looks best to you. You will then come to a page that allows you to change colors on the cover, change the position for parts of your cover and change fonts for the text. However, these positions are fixed by KDP, so you can't edit the locations within the arrangements they provide. Please note that you are now on step two of the process, called "Style and Edit." You'll see that on the top left of your page. You can click on Step One, "Choose Design," at any time by clicking on those words at the top left of your screen. You can also change whatever picture you chose by placing the cursor in the upper left of the cover image on your screen and clicking. Doing so will bring up a dialog box on the right side of your screen in which you can click on "Choose a new cover image."

In step two, Style and Edit, if you picked a cover template with colors and not an image that fills the entire cover, front to back, you can lightly click on the first icon on the lower left of the screen, which will say "choose colors" if you hover your mouse over it. You can pick one of the pre-prepared color schemes that show there, or you can customize them under "Custom Colors" at the lower-left corner of the screen. You can adjust both the primary color and secondary color separately. Just click on each and then pick a color from the checkerboard array that comes up. Please note that if you choose to use a pre-prepared color scheme, you can find more choices by clicking on the arrow to the right of the displayed choices.

To the right of the "Choose Colors" icon in the "Choose Layout" icon. If you click on this icon, you will see several choices of layout for your cover. For example, one layout might put your author's name above the title, and another one put your name at the bottom. There will also be options that move items on the back cover around, such as the book description, your author's picture, and the placeholder where the cover creator will insert your book's barcode based on your book's ISBN. Choose the layout that looks best to you, but remember that you have to go with one of these options - you can't click and drag content around.

Then click on each section on the back cover and fill them in. You can cut and paste in text from elsewhere or type right into the cover itself in the places provided. If your book is one hundred pages or more, you will see an area between your book's front and back covers called the spine. The title of your book and the author's name will appear on the spine printed vertically. Try to make sure that your book is at least one hundred pages to have a printed spine. Without a printed spine, no one will be able to recognize your book on a bookshelf. That fact can cause problems getting your book properly shelved or even purchased by bookstores and libraries. If you are close to but short of one hundred pages, think of adding some blank pages for notes in the back of the book or adding a bibliography or epilogue.

Please note that long titles may not fit on the spine even if your book is one hundred or more pages. If your title is too long, you will need to decide whether you will primarily sell the book through Amazon or other on-online distributors where people search for books using search words, or whether you will sell your book primarily through brick and mortar book stores where people peruse the shelves looking for titles of interest on the spines of the books they see on the shelves. Clearly, longer titles with more search words on the spine work best for online sales. So, if you must make a choice about whether to keep your longer title and give up having the title shown on the spine, make that decision by the primary marketplace for your book.

You will see that the cover creator will pre-load your title, subtitle, and author name but will not automatically fill in anything on the back cover. There will almost always be a place for the book description on the back cover. This description is very important, so work on it a bit. The book description will be a major marketing tool for your book. If people find your book on Amazon, they can read the cover. Often, purchase decisions are based solely on what that description says. Note that the spine of the book usually does not show on Amazon. However, purchasers can view the front and back cover of the book.

Some cover designs will have a place for you to put testimonials from others raving about your book. Suppose the design you've chosen does not have a separate section for those reviews. In that case, you can place them in the book description section along with a shorter explanation of your book.

You may also have chosen a design with a place for your publisher's logo. If you have an imprinted logo, you can upload it and put it here. If you don't have an imprinted logo and you've chosen a cover style with a place for one, just leave that place blank, and it won't show on the finished product. However, adding a publisher logo will make your book look more professional, and it will look less like you've self-published it. I've noticed that most of the KDP cover layouts no longer have a place for a publisher logo. I'm not sure I like that trend. I always think the book looks more professional if you add a publisher logo to the back cover. In fact, if you can add it to the spine, so much the better. At this point, the KDP cover creator doesn't appear to allow for putting your logo on the spine. However, if you have your cover custom designed, please make sure that you put a logo on the back and on the spine.

You can adjust the wording of your title and subtitle since the creator software actually allows you to edit it. However, your title should match what you have entered in that field in your KDP account, what you've entered for the ISBN in your MyIdentifiers Bowker account, and what you submitted to the Library of Congress when you requested your PCN.

Once you have entered all the necessary text, you can then move to the third and final icon, "Choose Fonts." Here you can pick the font in which you would like the title, subtitle, and author's name to appear. Keep clicking the arrow to the left of the choices to see all the choices. You can keep trying different font choices until you are satisfied. Once you choose your font combinations, make a note of which font you chose. You may want to use that font for the back-cover text as well. To do so, hover the mouse over the text on the back cover. Then, click there. You will see a line appear to the far left on your font choices for that text. Pick the font you want to use for that portion of your back-cover text. Use the same font for all your back-cover text to look consistent. When choosing fonts, bear in mind the subject of your book. Some fonts evoke images of certain periods or certain professions. So, pick the one you feel matches the flavor of your book.

At any time during the creative process, you can click on the gold "Preview" button at the bottom of your screen to see what your cover will look like with the options you've chosen. You can leave out certain items. Then, they simply won't appear on your final cover. So, for example, if you don't want your picture on the back, just don't upload one. If you don't want an author description, just leave that placeholder blank. You will need a barcode linked to your ISBN to sell the book, so you will find that you will not be able to delete that placeholder box. KDP will automatically fill that box for you. There is no need to purchase bar codes from Bowker.

If you've chosen a design with a place for your author photo on the back cover, you should upload a picture here. It's not absolutely necessary, but it's advisable to have an author picture on the back cover. It makes your book look more professional and lets people feel invested in your own work.

If you don't have a professional headshot, I recommend you get one. I had mine taken by Sean Kenney of www.bizphoto.com. I don't normally recommend services in my books, but Sean is really good at what he does, and he's really reasonable. If you are in the Los Angeles area or lucky enough to live somewhere he may be traveling for a shoot; you can use his services. He's the only person who can take a good picture of me, and he's a really nice guy with his own book out. It's always nice to support a fellow author. He was a model and actor himself before he switched to photography. He's taken the first headshots for lots of Hollywood stars. If you are sticking with the goal of publishing for free, then ask a friend who is good at photography to take your picture for you. Please don't try and take a selfie. This picture will help sell your book . . . or not. It needs to be a good one.

To upload your author photo, click on the space provided for your author photo, and then click on the box entitled "From my computer." You will see here that you also can skip this step if you really don't want your picture in the book. If you click the "From my computer" option, you can then locate the picture on your computer and upload it into your back cover.

I never used to like putting my picture on books, web pages, or business cards. Then, Sean took a great headshot of me, and I changed my mind. I now use that picture as branding. I offer a lot of different services, but my picture holds them all together and begins to create an expectation for a high level of quality of things associated with that picture.

We've become a society that largely looks for images. I've even seen them next to my hotel room numbers, which I frankly think is going too far, especially when the picture of a pasture next to my room number does not at all accurately represent what lies behind the door. Some of today's images seem irrelevant. However, your picture as the author of your book is relevant.

If you plan to use your book in any way to enhance your business, then you really should have an author picture on your book. Suppose your book is your only product, and you don't intend on attending any speaking engagements or doing any book signings. In that case, it probably doesn't hurt much to leave it off. Ultimately, that decision is up to you.

If you want to leave the cover creator before you've filled in everything on the cover, you can just click the "Save" button at the bottom of the screen at any time. Then you can exit and come back later to pick back up where you left off. You can also choose to start over from scratch any time you wish by using the grey "Start Over" button located at the bottom left of your screen.

Even if you've submitted your cover by pressing the "Save and Submit" button that will appear at the bottom of your screen when you are in preview mode, KDP allows you to go back in and change your cover. Again, you can make minor modifications to the cover because you are using a print-on-demand printer even after the first book has been printed. However, I recommend you get it as close to perfect as possible before submitting your book. Just know that if you find a mistake or typo after submitting the book, you can always go back and change it, save it and have it show up corrected for the next book printed. That's the beauty of on-demand printing!

While the KDP cover creator is the easiest way to create a free cover for your book, as we discussed, it also offers limited flexibility and customization. Now that many books have been published through KDP and its predecessor, CreateSpace, it's become easier to spot books designed with the cover creator, as there are only so many templates to choose from. So, the cover creator has begun to be something of a giveaway of a self-published product. Suppose you have a bit more time on your hands, some artistic ability, and a clear vision for what you want your cover to look like. In that case, there are ways to create your own completely customized cover for free, which KDP will still accept for printing.

The first step in creating your own cover without the aid of a cover creator is to figure out the exact dimensions of your cover. For this reason, do not start to create your custom cover until your page count is set. In the old days, CreateSpace had a calculator where you could enter your trim size and page count, and it would tell you the dimensions of your cover. KDP no longer offers this feature, but enterprising individuals have created a workaround. The marvelous non-profit website bookow.com offers a free cover size calculator, template generator, and even barcodes. While these products are all free, they do ask for donations. To use this service, go to www.bookow.com. Then, click "CreateSpace/KDP cover template generator." Enter your book information on the left. If you want a free ISBN barcode to come with your template, enter the ISBN information here; otherwise, leave it blank. The barcode generator further up on the page can create a barcode that you can separately paste onto your cover as an image if you prefer.

Books will email you your cover template in PDF form or several other formats customized to various design software programs. The white portion of the template corresponds to the "live area"—the part that you can fill with images, text, etc. The red part corresponds to material that will either be trimmed away or is bent over the spine. You can still fill this area with color but do not put images in this part.

Once you have this template, the choice is yours as to which software to use to create your full custom cover. Some of the common, high-powered design software options used by professionals include Adobe Photoshop and InDesign. These are both very powerful software tools that will allow you to create professional-looking, high-quality covers - but they have steep learning curves and are expensive. If you don't have these software programs or don't know how to use them and want to create your cover for free, there's another option - PowerPoint! Believe it or not, you can actually create a very convincing cover just using the basic tools provided in PowerPoint as part of the common Microsoft Office suite.

If you look at your Bookow template, it will tell you the overall template size, including bleed- which will be slightly larger than your actual trim size. Note that this overall length and width include the front, the back, and the spine. Open PowerPoint and create a new presentation. Change to a blank slide by choosing that option under "layout" in the upper left-hand corner of the tool ribbon. Once you have a single blank slide in your presentation, change the size of the slide to correspond to the size of the template. To do this, navigate to the "Design" tab, then select "slide size" on the far right. The drop-down menu will give you the option "custom slide size." Select this option, then manually enter the size of the template. Your slide will now be the correct size. Insert the template as an image (Books will send you a . PNG file, which will work for this application).

Now that your template is inserted, you can begin to work. Before you set about creating your cover, consider your design. Your cover design is very important, especially for online sales - often, the cover is only seen by potential customers as a small thumbnail image on a screen. So, your cover needs to be simple and striking enough that it will convey the message in a thumbnail, but it should also be detailed enough that it conveys something about the nature of your book. Nonfiction covers are, in some ways, easier than fiction covers, but they still require care in their design.

When choosing fonts for your cover, as mentioned above, try to pick fonts that match your genre and plotline. The beauty of creating your own custom cover is that you can choose any font you want- you aren't limited by the choices given to you by the KDP cover creator. For an even more professional look, tie all of your font choices together-make sure your exterior title font matches the font, or at least the font family, of your interior title page, chapter headings, etc.

Eventually, your content will cover the entire template. So, decide what your overall background color will be, then create a rectangular shape that covers the covering cover. The rectangular shape can be found under the "Home" tab in a large box with a variety of different shape choices. Once you've created the rectangle, you'll notice that the "shape fill" button is now available to use. After clicking on that button, you will be able to choose the color you want and any special effects like gradients. Note, however, that gradients and other skeuomorphic graphics (that is, graphics that try to create "3D" effects) are now considered dated. Once you've set your background rectangle color, right-click on the rectangle. Hit the button "send to back." Your rectangle will disappear. Don't be alarmed! You'll get it back later, but for now, you still need the template.

The next step is to create clear rectangles around all of the white portions of the template- the "live area." To do this, simply create rectangles around these areas (you will need three- one for the front and back and one for the spine). You'll also need another one for the yellow box that indicates where the barcode should go - the barcode location is standardized, so you shouldn't try to place your barcode somewhere else. When you initially create a rectangle shape (or any other shape) in PowerPoint, it will automatically fill blue. To get the rectangle to be transparent, click on "shape fill" and then "no fill." Now, the rectangle is just an outline. These four rectangles will serve as guidelines for you as you insert the rest of your content to your cover, and you will delete them at the end before you upload the cover.

Now that all of the important geometry of the template has been duplicated, you can send it to the back the same way you did with the color rectangle above. Note that the "send to back" feature puts the object behind everything else. So, if something was previously sent to the back, that object will no longer be in the back since something has now been sent behind it. So, your color rectangle will reappear, but it will still be behind your guideline rectangles, so you will still be able to see them.

Conversely, you can also bring an object to the front (which requires you to be able to select the object - so if it's totally blocked by something else, you'll have to move the object that's blocking it physically.) If you don't want to send something totally to the back but want it to move backward, hover over the arrow pointing to the right on the "send to back" button. "Send backward" will appear. This will send it back one layer. Essentially, every time you add an image or object, this creates a new layer. So, if you've inserted ten new objects and you want the object in question to be behind all of them, you will have to hit "send back

ward" ten times. This tool is useful if you want to superimpose images, like emblems, over other images. The "bring forward" feature works in reverse. It is accessible the same way, except by hovering over the "bring to front" button.

Now you can start to create your cover in earnest! If you have an illustration you would like to use for your cover (that you have the rights to), you can insert it as an image and completely control its position, size, etc. - this method is much more flexible than the cover creator.

Alternatively, you can even create basic illustrations using PowerPoint shapes. These are especially effective for illustrations or graphics solid in color - which is actually in fashion these days. Simple, graphic covers with bold colors and silhouetted images are actually more popular in many cases than more traditional covers with intricate artwork or paintings like you might find on a 1990s fantasy novel. At the end of the day, though, it's your cover, and it should look the way you want.

Sometimes, you want a shape for an illustration (or an ornament, like an emblem or symbol) that doesn't neatly match a regular shape, like a circle or polygon. PowerPoint can accommodate this. You can create complex curves using the "curve" tool, which looks like an upside-down "u," or perhaps a hill. It's found within the extensive selection of shapes, accessible either through the "Home" or "Insert" tabs. To create a complex curve, place control points by clicking your mouse. The curve will automatically be created to follow these points. The fewer points, the smoother the curve. After you create the curve, you can edit the positions of the points by right-clicking on the curve and selecting "edit points." This will let you click and drag the point. Another powerful shape tool is the "freeform" shape. This looks like an irregular, squiggly shape in the menu. You can place an unlimited number of points to define the edges of your shape. Once your return to the original point and place your final point to connect to it, your shape will fill with color. You can then adjust the color the same way you would for a normal shape. If you right-click on the shape and hit "format shape," a dialogue box will open to the right with more options for special effects. You can fill the shape with a pattern, texture, image or even make it transparent or semi-transparent. I've found this freeform shape tool very useful. I've used it to create maps in books and even create basic artwork of aircraft and ships.

If you've created a freeform shape and want to resize it while maintaining its proportions and aspect ratio, click on the shape and drag one of the corner handles while holding down the shift key. Similarly, if you are drawing lines, or want part of your freeform shape to either be horizontal, vertical or at an exact forty-five-degree angle, hold the shift key down while drawing the line. Sometimes, lines will also snap to control points on nearby curves. This is great if it's what you actually want and really annoying if it isn't.

If you have a group of images or shapes that you want to move together, there are a couple of ways to accomplish this. If you hold down the control key while individually selecting each shape with a mouse, then you've selected them all and can move them together. Similarly, suppose you hold down your mouse and drag. In that case, this creates a box that will select any object that is fully within the box (if a shape goes outside of the box at all, even a little bit, it will not be selected), and you can move them together.

Note that for all shapes, you can also adjust the outline. If you hit the "shape outline" button, you can change the color. You can also change the weight- this means you can adjust how thick the line is. The "dashes" option gives you the ability to make the border a dotted line, with several different options. You can also select "no outline," which will mean that the shape will not have a distinct outline, just the fill color if you've specified one.

If you have images with backgrounds that you don't want to show (such as a logo with a white background or a picture of a building where you just want the building, not the surroundings), two tools can help you. If the background is a solid color, double click on the image. A new set of tools will appear on the top ribbon. Click on "color" near the left, then hit "set transparent color" near the bottom of the dialogue box. This will enable you to pick one color that will then become transparent.

Note that it's better to have a transparent background than to try to color-match a background to its surroundings when creating an image - even if you think you've picked the right color, it might still show up as slightly different when printing (this happened to me once). This tool only works if you have a vector image where the background color is consistent. If this is not the case, double click the image and, this time, select the "remove background" option. PowerPoint will guess which part of the image you want to keep and which is the background. It will probably guess wrong. Fortunately, you can use the "mark areas to keep" tool to draw lines through the parts you want and the "mark areas to remove" tool to draw lines through the part that you don't want.

I've found that a little bit goes a long way with these tools. Rather than creating extremely detailed demarcations, you can literally draw a slash through the area you want, and PowerPoint will eventually figure it out. You can also adjust the boundaries of the part of the picture that PowerPoint is examining, essentially the reverse of a cropping operation.

PowerPoint also offers a number of edits and special effects for images. If you hit "recolor," it will give you a limited number of options for you to recolor your image. "Corrections" will help you adjust the brightness, sharpness, and contrast. "Artistic effects" gives you the ability to make a photograph look more like a painting, as well as several other features which range from the useful to the bizarre, such as an option to make your image look like it's covered in plastic wrap. To bring any of these options up, double click the image or right-click and select "format picture." Sometimes, if you try to do too many things at once (like adjust the contrast *and* add an artistic effect), PowerPoint will freak out and grey out the options. To get around this, simply click somewhere else on the screen (to deselect the image) and then double-click the image again.

Creating your own cover also gives you the ability to place your imprint logo on the back of the book and on the spine. Publisher and imprint logos usually don't go on the front cover of the book. A note on spine design—the live area of the spine, where you can put content, is flanked on either side by strips of non-live area, where the cover paper is folded over the corners of the book to transition to the front and back covers. While the spine is usually centered properly, you have next to no control over KDP's printing quality control, and it's possible that your spine won't be exactly centered relative to the edges of the book. To make this less obvious, simply carry a consistent color from the front cover, through the folding areas, the live area of the spine, and the back cover. This way, it will be harder to tell if the spine isn't centered than if the spine's live area was a different color.

Much like images, PowerPoint offers a number of ways to edit text. Select the text you want to edit, then click "format text effects." The resulting toolbar on the right of the screen will give the ability to add features like shadows, textures within letters, glow effects, etc. I actually don't recommend doing too many special effects with letters. It can make your text harder to read, and opposite to the desired effect, it can actually make your work look dated and homemade. However, judicious use of solid color can go a long way. Suppose you would like to add a different-colored border around letters. In that case, you can do that by selecting the "text outline" option under the "A" on the left side of the toolbar. It

works much like the shape outlines discussed earlier.

One of the last steps is to add a barcode to the back of the book, in the position marked by the Books template. Once again, you do *not* need to pay Bowker's $2 5 fees for a barcode. If you're using a KDP template, KDP will automatically create the barcode for you. If making your own cover, bookow.com can create a barcode for you. Go back to the Books website and simply enter the ISBN and the retail price.

If you want to leave the price flexible, enter "90000" in the entry box for the price. Don't leave it blank, as it's best to include this information in the barcode even if you don't have a set retail price. Leaving the retail price unset allows you to run discount promotional events, for example. When the barcode is generated (much like the template, Bookow will email it to you in a couple of usable formats), you will notice it includes two main sections -

the primary barcode on the left, which includes the ISBN information, and a shorter barcode on the right, which includes the price information. If you actually do enter a price, you'll notice that the numbers along the top of this barcode will correspond to the dollar amount of your price.

Once you've received the barcode from Books, paste it in as an image and maneuver it over the guide rectangle that you created for the barcode at the beginning of this process. If you've used Bookow's services, don't forget to thank them with a donation.

Once you've added everything to your cover, delete the guide rectangles, or send them to the back if you think you might need them later. You're now ready to export your cover into a usable format. PDF works well for KDP. To do this, simply save your presentation as a PDF. PowerPoint may ask you which slides to export (presumably, there's only one in your presentation) - just pick "current slide." You can then upload this PDF as your cover in KDP.

Remember that you will probably want a JPEG image of just your front cover to use as a thumbnail on Amazon. You will also need it for the cover of your Kindle version since ebooks don't have a back cover or a spine.

The Final Review

The final tab in the set-up process is the "Paperback Rights and Pricing" tab. The first box under this tab is "Territories." Again, for ease of instruction, I will lay out the steps to take under each box label.

The Territories Box:

Click on the bubble for "All territories (worldwide rights)" as long as you or the author for whom you are publishing actually wrote the book and no one else had bought any of the rights associated with the book.

The Pricing and Royalty Box:

Next, in the "Pricing and Royalty" box, choose Amazon.com as the primary marketplace if you are located in the USA. If you are working in a different country, use the arrows in that box to take you to the correct Amazon division for that country.

Next to the Primary Marketplace box is the list price box. Before you start entering numbers in that box, click the empty box next to the words "Expanded Distribution," which you will find right under the list price box. Checking this box will enable your book to be sold to bookstores, libraries and to show up for sale on other online retailer sites such as barnesandnoble.com. Of course, you will need to convince libraries and bookstores to carry the book yourself. KDP will not do that kind of marketing for you. However, other online retailers will generally, automatically pick it up for sale, and Google Books should give you

a list of online retailers who carry your book.

Before going back and entering a price for your book in the "List Price" box, go to www.Amazon.com and do a search for books like yours. Do not price your book higher than the average for other books of your type. If you are a new author without a following, you may wish to price your book below the average to attract cost-conscious buyers.

Once you've settled on a beginning price to try, enter it in the list price box. When you do, you will see that KDP then calculates the royalties you will receive per purchase for all the different distribution channels you have selected. You can then play with the retail price until you are happy with the royalties you will receive.

Please note that it is possible to price the book so low that you won't receive a royalty through expanded distribution. If that happens, I recommend raising the price until you get at least five cents from a sale through expanded distribution.

Also, note that if you list your book for sale on Amazon, Amazon will likely discount the price. However, the royalty you establish through your KDP account will not change. In other words, Amazon eats the amount of the discount. So, the only way that discount negatively affects you is that it will be hard for you to sell the book yourself for more than the price people will pay for it on Amazon. However, remember that Amazon also adds on shipping and taxes. So, unless you're selling to an Amazon Prime member and you won't pay the sales tax for the purchaser, you may still have a price advantage over even a discounted Amazon sale.

Below the words "Expanded Distribution," you'll see the blue words, "6 other marketplaces." Click the arrow next to those blue words. KDP will open a list of other Amazon sales channels outside the United States and price your book in the currency for each of those countries. KDP has set the price in those countries based on the US dollar price you entered for your book. You can override this default if you wish to specify custom prices for each country. Just put the cursor in the box you want to change and change the price. If you later decide you want to let KDP adjust the price in other countries based on the US dollar retail price, then just click on the words under each box, "Based on Amazon.com." Doing so will revert the foreign prices to the foreign exchange equivalent of the

US price. If you are writing and publishing your book in a country other than the United States, I would assume that the primary retail price box would default to whatever country your working in and that the United States would then be listed as one of the six other marketplaces.

Once you've entered your prices, you will see that you will receive a sixty percent royalty on sales made through an Amazon site and forty percent on those sold through other channels. Amazon first deducts the cost of printing the book and then calculates the royalty on the amount leftover.

Make sure to save your pricing choices by clicking on "Save as Draft" at the very bottom of this tab page. Please do this before going on to any more sections under this tab.

The Terms and Conditions Box:

You will also see a box labeled "Terms and Conditions." Here, KDP tells you that it could take up to seventy-two hours for your book to appear on Amazon once you click the gold "Publish Your Paperback Book" button and that by clicking that button, you agree to all their terms. Again, those terms are not negotiable, so you either publish your book under those terms, or you don't publish your book with KDP. However, don't click the "Publish" button quite yet.

The Proof Copies Box:

The line right above the "Publish" button allows you to order a proof copy of your book. I strongly suggest you do that, receive the proof copy, and review it before pushing the publish button. The cost of your a proof copy will be the cost to print your book, plus shipping and possibly sales tax as mentioned above.

It is hard to wait for that copy to show up in the mail, but the chances are very high that there is something wrong with the text or layout that you couldn't see in the electronic preview version. I once received a proof copy of a book I thought was perfect, only to find that I had left off the period from the book's last sentence! I couldn't really see that in the electronic previewer, but it was obvious when I thumbed through the actual, physical proof copy. So, order the proof copy and review it before pushing the "Publish" button.

Note that proof copies now come with a large ribbon printed around the cover to prohibit resale. Something we've also experienced on some proof copies is a non-centering of the spine content, which for some reason naturally worked itself out in the "production" version.

Also, before you hit the "Publish" button, go back through all the boxes in each of the tabs for your book. Make sure that all of the settings are correct since you will not be able to change some of these. After your book is "published." Pay particular attention to the title of your book and the book trim size you have chosen. Make sure that they exactly match what you intend. You can easily miss a word or click on the wrong trim size. So, double-check those entries before you click on "Publish."

The Publishing Your Paperback Button:

Once you have received the proof copy and are happy with it, go ahead and click the Publish button! Congratulations! You are now a published author. If you use your own ISBN for your book, then double congratulations! You are now also a publisher! Remember that although the button is entitled: "Publish Your Paperback Book," you are just completing the set-up process for your printer, KDP. Your book should be ready for you to order copies to sell yourself, and it should appear for sale on Amazon within seventy-two hours. Although you have now completed the ten easy steps to self-publishing, you will want to read the following appendixes, which explain that important follow-up you should consider as well and publish a Kindle edition of your book.

We wish you many happy sales and a real sense of accomplishment!

Setting The Right Price

The final tab in the set-up process is the "Paperback Rights and Pricing" tab. The first box under this tab is "Territories." Again, for ease of instruction, I will lay out the steps to take under each box label.

The Territories Box:

Click on the bubble for "All territories (worldwide rights)" as long as you or the author for whom you are publishing actually wrote the book and no one else had bought any of the rights associated with the book.

The Pricing and Royalty Box:

Next, in the "Pricing and Royalty" box, choose Amazon.com as the primary marketplace if you are located in the USA. If you are working in a different country, use the arrows in that box to take you to the correct Amazon division for that country.

Next to the Primary Marketplace box is the list price box. Before you start entering numbers in that box, click the empty box next to the words "Expanded Distribution," which you will find right under the list price box. Checking this box will enable your book to be sold to bookstores, libraries and to show up for sale on other online retailer sites such as barnesandnoble.com. Of course, you will need to convince libraries and bookstores to carry the book yourself. KDP will not do that kind of marketing for you. However, other online retailers will generally, automatically pick it up for sale, and Google Books should give you a list of online retailers who carry your book.

Before going back and entering a price for your book in the "List Price" box, go to www.Amazon.com and search for books like yours. Do not price your book higher than the average for other books of your type. If you are a new author without a following, you may wish to price your book below the average to attract cost-conscious buyers.

Once you've settled on a beginning price to try, enter it in the list price box. When you do, you will see that KDP then calculates the royalties you will receive per purchase for all the different distribution channels you have selected. You can then play with the retail price until you are happy with the royalties you will receive.

Please note that it is possible to price the book so low that you won't receive a royalty through expanded distribution. If that happens, I recommend raising the price until you get at least five cents from a sale through expanded distribution.

Also, note that if you list your book for sale on Amazon, Amazon will likely discount the price. However, the royalty you establish through your KDP account will not change. In other words, Amazon eats the amount of the discount. So, the only way that discount negatively affects you is that it will be hard for you to sell the book yourself for more than the price people will pay for it on Amazon. However, remember that Amazon also adds on shipping and taxes. So, unless you're selling to an Amazon Prime member and you won't pay the sales tax for the purchaser, you may still have a price advantage over even a discounted Amazon sale.

Below the words "Expanded Distribution," you'll see the blue words, "6 other marketplaces." Click the arrow next to those blue words. KDP will open a list of other Amazon sales channels outside the United States and price your book in the currency for each of those countries. KDP has set the price in those countries based on the US dollar price you entered for your book. You can override this default if you wish to specify custom prices for each country. Just put the cursor in the box you want to change and change the price. If you later decide you want to let KDP adjust the price in other countries based on the US dollar retail price, then just click on the words under each box, "Based on Amazon.com." Doing so will revert the foreign prices to the foreign exchange equivalent of the

US price. If you are writing and publishing your book in a country other than the United States, I would assume that the primary retail price box would default to whatever country you're working in and that the United States would then be listed as one of the six other marketplaces.

Once you've entered your prices, you will see that you will receive a sixty percent royalty on sales made through an Amazon site and forty percent on those sold through other channels. Amazon first deducts the cost of printing the book and then calculates the royalty on the amount leftover.

Make sure to save your pricing choices by clicking on "Save as Draft" at the very bottom of this tab page. Please do this before going on to any more sections under this tab.

The Terms and Conditions Box:

You will also see a box labeled "Terms and Conditions." Here, KDP tells you that it could take up to seventy-two hours for your book to appear on Amazon once you click the gold "Publish Your Paperback Book" button and that by clicking that button, you agree to all their terms. Again, those terms are not negotiable, so you either publish your book under those terms, or you don't publish your book with KDP. However, don't click the "Publish" button quite yet.

The Proof Copies Box:

The line right above the "Publish" button allows you to order a proof copy of your book. I strongly suggest you do that, receive the proof copy, and review it before pushing the publish button. The cost of your a proof copy will be the cost to print your book, plus shipping and possibly sales tax as mentioned above.

It is hard to wait for that copy to show up in the mail, but the chances are very high that there is something wrong with the text or layout that you couldn't see in the electronic preview version. I once received a proof copy of a book I thought was perfect, only to find that I had left off the period from the book's last sentence! I couldn't really see that in the electronic previewer, but it was obvious when I thumbed through the actual, physical proof copy. So, order the proof copy and review it before pushing the "Publish" button.

Note that proof copies now come with a large ribbon printed around the cover to prohibit resale. Something we've also experienced on some proof copies is a non-centering of the spine content, which for some reason naturally worked itself out in the "production" version.

Also, before you hit the "Publish" button, go back through all the boxes in each of the tabs for your book. Make sure that all of the settings are correct since you will not be able to change some of these. After your book is "published." Pay particular attention to the title of your book and the book trim size you have chosen. Make sure that they exactly match what you intend. You can easily miss a word or click on the wrong trim size. So, double-check those entries before you click on "Publish."

The Publishing Your Paperback Button:

Once you have received the proof copy and are happy with it, go ahead and click the Publish button! Congratulations! You are now a published author. If you use your own ISBN for your book, then double congratulations! You are now also a publisher! Remember that although the button is entitled: "Publish Your Paperback Book," you are just completing the set-up process for your printer, KDP. Your book should be ready for you to order copies to sell yourself, and it should appear for sale on Amazon within seventy-two hours. Although you have now completed the ten easy steps to self-publishing, you will want to read the following appendixes, which explain that important follow-up you should consider as well and publish a Kindle edition of your book.

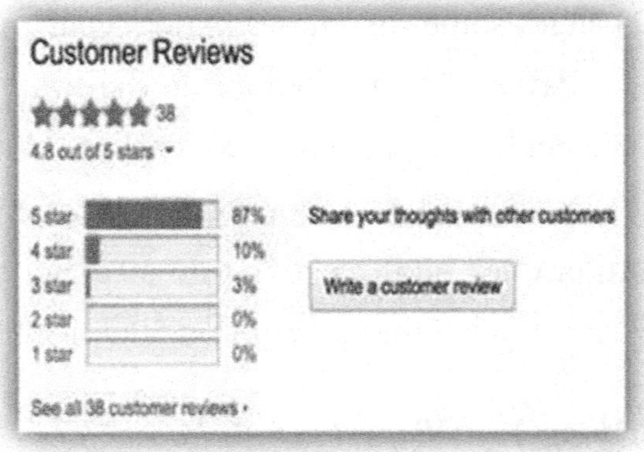

Please Leave a Review!

I would be incredibly thankful, if you could take just 60 seconds to write a brief review on Amazon. Even it's just a few Sentence!

Conclusion

So, as you can clearly tell now from going through the book, there are lots of legitimate ways through which you can start earning money online and make a living. I mean, who doesn't want to leave their 9-5 repetitive job for a job where they have complete freedom over their time, right?

If you work online, you will be able to take full control of your time and your life as well! You can work whenever you want, wherever you want, all depending on your personal preference and the level of workload that you want to take.

I'm not sure how many times I need to emphasize this, but I want you to understand that YOU WILL NOT BECOME A BILLIONAIRE OVERNIGHT. No, you will not be able to earn seven figures in a single day. Unless you win the lottery or a bet, there is no magic to achieving that. You will eventually become a millionaire. If you follow the proper procedures, you will not be able to earn seven figures all at once. You will be less likely to fall into the hands of fraudsters if you are aware of this.

While the internet can be a great way to make money and make connections, it can also be difficult to determine the right kind of product or service to provide and find the right people who are willing to pay for what you are offering. This, however, is not a problem without a solution, and once you grasp how things are done, you can be certain of making progress.

Every product and service that man requires is most likely already available online. You could simply be joining a competition queue, which is a common issue for many people, particularly newcomers to the labor market. However, the key to your survival in this sea of fishes is to stand out and add at least one or two plot twists to what people are already accustomed to.

I hope this book has been helpful to you along the way and that your relationships are improving. Thousands, if not millions, of people out there trying to get on the hype train of making money online, but only a few of them make it out alive due to the sheer dedication and willpower required to make things work!

When starting your own online business, one of the most difficult challenges is establishing a brand. This challenge entails processing payments, designing a logo, developing a website, handling customer complaints, and providing the best customer care services possible. These issues frequently act as a hindrance or impediment to achieving desired outcomes. For some, the cost of hiring others who can handle these matters is prohibitively expensive. Regardless of the drawbacks, there is always a way to get around this obstacle, and once you do, you are one step closer to your expected earnings.

Your first attempts might not work, but the idea is never to give up! Keep trying the different tactics found in this book, and I promise you that eventually, you will be able to overcome your woes and develop a highly successful career in making money online through one means or another.

As a final request, I would encourage you to leave feedback for my book on Amazon to find this book helpful and effective. It would mean a lot to me and inspire me to produce more high-quality content for you in the future!

Stay Safe and Stay Blessed.

Made in the USA
Monee, IL
05 November 2026

67845320R00291